The International Library of Environmental, Agricultural and Food Ethics

Volume 31

Series Editors

Michiel Korthals, Wageningen University, Wageningen, The Netherlands

Paul B. Thompson, Michigan State University, East Lansing, USA

The ethics of food and agriculture is confronted with enormous challenges. Scientific developments in the food sciences promise to be dramatic; the concept of life sciences, that comprises the integral connection between the biological sciences, the medical sciences and the agricultural sciences, got a broad start with the genetic revolution. In the mean time, society, i.e., consumers, producers, farmers, policymakers, etc, raised lots of intriguing questions about the implications and presuppositions of this revolution, taking into account not only scientific developments, but societal as well. If so many things with respect to food and our food diet will change, will our food still be safe? Will it be produced under animal friendly conditions of husbandry and what will our definition of animal welfare be under these conditions? Will food production be sustainable and environmentally healthy? Will production consider the interest of the worst off and the small farmers? How will globalisation and liberalization of markets influence local and regional food production and consumption patterns? How will all these developments influence the rural areas and what values and policies are ethically sound?

All these questions raise fundamental and broad ethical issues and require enormous ethical theorizing to be approached fruitfully. Ethical reflection on criteria of animal welfare, sustainability, liveability of the rural areas, biotechnology, policies and all the interconnections is inevitable.

Library of Environmental, Agricultural and Food Ethics contributes to a sound, pluralistic and argumentative food and agricultural ethics. It brings together the most important and relevant voices in the field; by providing a platform for theoretical and practical contributors with respect to research and education on all levels.

More information about this series at http://www.springer.com/series/6215

Cristian Timmermann

Social Justice and Agricultural Innovation

Springer

Cristian Timmermann
Interdisciplinary Center for Studies
in Bioethics
University of Chile
Santiago, Chile

ISSN 1570-3010 ISSN 2215-1737 (electronic)
The International Library of Environmental, Agricultural and Food Ethics
ISBN 978-3-030-56192-5 ISBN 978-3-030-56193-2 (eBook)
https://doi.org/10.1007/978-3-030-56193-2

This Springer imprint is published by the registered company Springer Nature Switzerland AG
The registered company address is: Gewerbestrasse 11, 6330 Cham, Switzerland

Para Paula y Esteban

Acknowledgements

I am strongly indebted to the many helpful questions, comments, remarks, complaints, and worries I have received, and want to express my gratitude to the many participants and organizers who have provided me with platforms to present earlier versions of this work.

To begin, I wish to thank the Colegio de Bioética, through an invitation by Gustavo Ortiz Millán, and the Division of Juridical Studies at CIDE in Mexico City for being the first audiences to help me shape my ideas and provide feedback on examining innovation as a (at that time) four-dimensional justice problem. In this regard, I also want to thank Michiel Korthals for suggesting me this book series and his numerous editorial recommendations. Later on, I had the opportunity to present the first outline of this book to audiences at the Social Justice Seminar of the Faculty of Law at the Universidad de Chile, thanks to a kind invitation from Pablo Aguayo, and to the group based around Brad Sherman, who accommodated me with great hospitality for a month at the T.C. Bernie School of Law at the University of Queensland, Australia, and at the V Congreso de la Asociación Chilena de Filosofía in Valdivia at the Universidad Austral de Chile. A much later version of this framework, after it grew to include six distinct forms of justice, was presented at the Hastings Center, United States. My discussions with the Center's researchers and the other visiting scholars during my month-long research stay were very fruitful. I am also grateful to an audience at the Brocher Foundation in Switzerland, where I discussed this framework as applied to the related problem of public health.

For specific sections of the book, I owe special thanks to Shuji Hisano, at the Graduate School of Economics at Kyoto University, for providing me with space to discuss the concept of agricultural justice in an interdisciplinary and intercultural context. Here I also want to thank Zoë Robaey for helping me to explore the relationship between Responsible Research and Innovation and theories of justice, and for setting up a workshop at Delft University to discuss these ideas, as well as for her written comments on the introduction. These discussions provided a very helpful basis to build on. Later on, the researchers involved in the Just Food project at the University of Jyväskylä in Finland provided me with an excellent opportunity to discuss in great detail a more elaborated version of this concept. I am grateful to Teea Kortetmäki for organizing the lecture and discussion.

Johan Olsthoorn was kind enough to organize a seminar on Chap. 4, on justice in exchange, at the Katholieke Universiteit Leuven, Belgium. The discussion during the seminar and his detailed comments were crucial for improving this chapter.

Earlier versions of Chap. 5, on distributive justice, were presented at a Law School Seminar organized by Eduardo Rivera López at the Universidad Torcuato di Tella, in Buenos Aires, and at the 5th Conference of the Latin American Association for Analytic Philosophy, in Villa de Leyva, Colombia. I benefited enormously from the feedback I received from the participants, and I owe a special thanks to Julian Reiss for his literature suggestions.

Some of the main ideas of Chap. 6, on contributive justice, were presented in Lyon at a Citizen Science conference. I am grateful to the organizers and the audience for encouraging me to think about the relationship between contributive justice and citizen science. A revised version of this paper was also presented at a workshop chaired by Zoë Robaey on Responsible Research and Innovation at the University of Tampere, Finland, and at a workshop organized by Brendan Coolsaet at the Université Catholique de Lille, France. The feedback I received there allowed me to revise some key sections of this chapter.

I also want to thank Alonso Donoso for inviting me to discuss the key environmental issues of innovation in relation to social justice at a Workshop on Justice and Non-Humans held at the Pontificia Universidad Católica de Chile. The main ideas discussed there were divided between Chaps. 7 and 9.

I had the opportunity to present early drafts of Chap. 7, on intergenerational justice, at the Yaoundé Seminar on Political Philosophy at the Université Catholique de l'Afrique Centrale, in Yaoundé, Cameroon, and at the 11th International Conference on Applied Ethics at Kyoto University. Here too, I am very indebted to the feedback I received, and I owe a special thanks to Tim Meijers and Dominic Roser for insightful discussion and some conceptual clarifications.

In relation to Chap. 9, I wish to thank Cristóbal Bellolio for organizing the workshop at the University of Manchester where I presented an early version, and Sanna Lehtinen for giving me an excellent opportunity to discuss a revised version at the University of Helsinki.

I also want to use this opportunity to thank the people at the Chaire Hoover d'éthique économique et sociale, at the Université Catholique de Louvain, for hosting me for 2 months. Their collegiality and advice were greatly appreciated in helping me through the final stages of the manuscript. I owe special thanks to Axel Gosseries for inviting me to the Chaire and for organizing a book-in-progress workshop, where I benefited from his feedback and the extensive commentaries provided by Eric Boot and Maxime Lambrecht. I also thank Christine Frison for her observations during the workshop.

Moreover, I learned a lot from the people at the Farming Systems Ecology group at Wageningen University about agroecology, agricultural extension, and their field research observations. Georges Félix, who introduced me to that group, also helped me to apply the concept of contributive justice to our earlier paper on agroecology and read over parts of the manuscript.

John Horden did an excellent job in English language editing and spotting some key mistakes in one of the final versions of the manuscript. I am very grateful for his work.

I owe special thanks to Fernando Lolas for believing in this project from the very beginning; assisting me with the funding application; introducing me to his network; and providing valuable help, comments, literature suggestions, and encouragement.

Unfortunately, I am to blame for all the remaining mistakes.

In addition, I want to thank for the great hospitality I received during different stages of the writing process in Santiago to my brother Esteban and his wife Janine, in Mexico City to Astrid, in Quillón to my mother and Juan Carlos, in Coyhaique to my sister Paula, in Wageningen to Georges and his housemates, and especially to my father for hosting me for such an extensive period in his home in the mountains of Quilaco. The times shared greatly improved the writing experience and quality. They also had nice pets.

Most importantly, I wish to thank all the residents and visitors of Chile for funding this project with their taxes, through my postdoctoral fellowship FONDECYT/CONICYT No. 3170068.

Contents

Chapter 1
The Multiple Dimensions of Social Justice Affected by Agricultural Innovation

Abstract This introductory chapter provides an overview of the different philosophical approaches to assessing technologies. It starts by discussing today's great challenges for providing food in a socially and environmentally sound manner. It continues with a brief overview of the different schools of post-World War II technology assessment. After that, it introduces the six-dimensional social justice framework used in this book and provides definitions of the main concepts.

Keywords Technology assessment · Responsible research and innovation · Ethics · Theories of justice

1.1 Introduction: Developments, Achievements and Failures

Food production is inherently embedded in a changing social and natural environment. Agricultural innovation and science policies have to embrace the fact that they are introduced into a system of complex interactions and factors: a changing climate and environment, migrating living organisms, drifting inorganic matter, fluctuating populations and varying social needs (McIntyre et al. 2009). These changes affect social and environmental interactions and thus require innovation systems to continuously (re)align with the demands of social justice. Food procurement is a major social justice challenge, as constantly transforming food systems have to be continuously assessed and readjusted to meet the demands of justice in all its dimensions.

Current as well as future food needs raise issues of justice. To start, it is crucial to assess how well a farming system meets humanity's immediate needs, that is, to what extent the system delivers food in sufficient quantities and adequacy, under conditions that can consistently satisfy demands around the world (Tittonell et al. 2016). We are currently very far from reaching from this target. Depending on how hunger is defined, it is estimated that one in every seven to nine people in the world does not have enough food to live a flourishing life (Lappé et al. 2013): a morally unacceptable state of affairs that demands urgent action. In terms of food adequacy, inadequate food supplies and diets lead to malnutrition, leaving people with insufficient vitamins and

C. Timmermann, *Social Justice and Agricultural Innovation*, The International Library of Environmental, Agricultural and Food Ethics 31,
https://doi.org/10.1007/978-3-030-56193-2_1

minerals to cover vital needs to allow full physiological functioning and development, sometimes even leading to blindness and reduced cognitive development (Ziegler 2011). In addition, we need to address the problems caused by obesity. Lack of access to fresh food, sedentary lifestyles and inadequate diets have caused an obesity epidemic that is having devastating effects on health and well-being (Lolas 2014). The commodification of wide areas of the global food sector has come with a massive increase in unsustainable and unhealthy diets (De Schutter 2017). Agriculture needs to deliver a sufficient quantity of food in wide enough variety to satisfy dietary needs in a culturally adequate manner.

Moreover, many food production sites are not socially sustainable, as they rely on huge numbers of migrant labourers, engage workers in conditions similar to slavery, and produce an insufficient work-harvest ratio (Loo 2014; Timmermann et al. 2018). Food production systems should not lead to avoidable health hazards, and this requires major work in reducing exposure to pathogens from animal husbandry and pesticides (Canavan et al. 2017). The agricultural sector also needs to be reformed in order to be socially sustainable in terms of protecting farm workers' health and rights, and provide decent standards of living in rural areas.

Food systems also need to be economically sustainable. Innovation should not lead to a situation where farmers become bankrupt on a massive scale, abandoning rural areas, which again destroys rural businesses and makes it difficult to maintain an adequate rural infrastructure (Thompson 2017). The resulting concentration of land in a few hands imperils food sovereignty and gives food producers a hard-to-control power over a basic good, thus jeopardizing food entitlement (Nyéléni Forum for Food Sovereignty 2007).

Since food is not only a need in the present, it is mandatory for food systems to be sustainable in the long-term, so that future generations can secure their food needs. Nowadays, supplying the world's population with food is already having an enormous environmental impact, in terms of greenhouse gas contributions, loss of biodiversity, pollution, and a strong dependency on fossil fuels for the production of fertilizers, food transportation and the operation of machinery (Tittonell 2013; Patel and Moore 2017). Since agricultural development in recent decades has tended to ignore the negative externalities of food production and underappreciate the multifunctionality of agriculture, major changes are needed in how we look at nature, its services and value (Pretty 2008; Altieri et al. 2017). It will take a major effort through multiple lines of action to ensure that food is produced and consumed not only in sufficient quantities and adequacy, but also in an environmentally sustainable manner (Springmann et al. 2018). We need to take seriously the idea that people in the future will also want to live on a planet that harbours an acceptable amount of wildlife (IPBES 2019).

To address these enormous social and environmental problems, proponents of both conventional and ecological agriculture claim that innovation is essential for addressing future food security, reducing our environmental footprint and improving rural livelihoods (Borlaug 2007; Tittonell 2013). Yet, especially since the early 1980s, partly due to the seminal work of Amartya Sen, scholars and policy-makers around the globe have become increasingly consciousness of the fact that hunger is not only a scientific problem, but also a social problem. The biggest famines in the last century have occurred in countries where enough food was available (Sen 1981). In fact, hunger is still widespread nowadays, even though the world produces more than enough food to feed us all (Tittonell et al. 2016). Amartya Sen's conclusions are that we need to reinforce democratic institutions, by widening participation in the development of new policies and increasing transparency (Sen 1981). This book builds on this perspective. Since food production needs to be increased to feed growing populations, efficient and transparent governance cannot be limited to the allocation of tangible resources for the production of food—such as water, land and fertilizers—but needs to be expanded to also include a just governance and production of intellectual resources. As advocates of food sovereignty assert, access to the means of production also requires access to technological and social innovations, and the development of scientific solutions tailored to addressing the needs of the poor (Nyéléni Forum for Food Sovereignty 2007; Via Campesina 1996). In order to meet these other needs, a wider participation in the design, development, use, choice and good governance of agricultural innovations is necessary (Pant 2019). Moreover, establishing the right to self-determination requires democratic decision-making regarding agricultural policies, which in turn requires improving political literacy among the rural population and restricting the power of corporate lobbyists to excessively interfere in the political arena (Wise 2019).

In order to facilitate a systematic assessment, we need to pay attention to three characteristics of agricultural innovation that are not always shared with other scientific and technological fields: instability, non-universality and necessity.

First, concerning instability, it must be noted that innovations are embedded in a changing social and natural environment. Natural organisms react to most types of innovation in agriculture (Goeschl and Swanson 2003). A standstill in innovation does not mean that we will retain our food production capabilities. We need to constantly adapt food production to changes in the environment, such as anthropogenic climate change, variations in precipitation and temperatures, the migration of parasitical and commensal species, soil deterioration, levels of carbon and nitrogen, and chemical pollution, among other factors (McMichael 2017; Cline 2007). Degraded farmlands need to be recovered, and adequate soil management measures need to be widely implemented to prevent future erosion and the loss of soil fertility (Tittonell 2016; El Mujtar et al. 2019). Rising temperatures, the international food and animal trade, and the mass migration and travel of people facilitate the propagation of pathogens. Many pathogens are becoming resistant to pesticides (Popp et al. 2013). Population growth demands that more food is produced on less land while reducing the environmental impact of agriculture (De Schutter 2017). Our natural environment and human populations have a dynamic character, and in order to maintain food production capacities we need to counter these environmental and social changes with a sufficiently fast advancement of agricultural science (Mazoyer and Roudart 2006). In the tropics, with a harsher climate, faster plant growth and weaker economic conditions, it may even take substantial innovation to reproduce the results of previous harvests (Richards 1993). Considering these factors, we should go beyond talking about the necessity of agricultural innovation, which we cannot deny if we want to survive as large societies, and instead concentrate on how we are to improve these innovation systems and expand the diffusion and uptake of innovations suited to local social and environmental conditions.

As for the second issue, non-universality, we need to keep in mind that much agricultural innovation is adapted to specific ecosystems, terrains, climate zones, cultural adequacy, and market preferences (Gliessman 2007). Innovations that work in one area or for one particular purpose may not function properly under other environmental and social conditions (McIntyre et al. 2009). Ideally, research should be carried out by decentralized research centres to identify site-specific needs and successes, distributed all over the world to account for geographic, climatic and social diversity. Weaker institutions need to be strengthened so that they can adequately serve local needs. New research facilities need to be established in places that are currently underserved by the scientific community.

Third, with respect to necessity, agricultural innovation is essential for securing the human right to food,[1] which therefore makes it mandatory under international agreements and several national constitutions (De Schutter 2009a). States, alone and in cooperation with others, have the obligation to facilitate and stimulate such research efforts (UN Committee on Economic Social and Cultural Rights 1999). This creates major limitations on how far innovators can pursue their economic interests without considering social needs.[2] Excessive rent-seeking in relation to agricultural innovations jeopardizes our ability to benefit from scientific advancement by securing the right to food.

In standard technology assessment, policy analysis and social research on agricultural innovation and food politics have concentrated their attention on two social justice issues: access to the fruits of innovation and the means of food production,

[1]See the Universal Declaration of Human Rights (1948) art. 25.1 "Everyone has the right to a standard of living adequate for the health and well-being of himself and of his family, including food, clothing, housing and medical care and necessary social services, and the right to security in the event of unemployment, sickness, disability, widowhood, old age or other lack of livelihood in circumstances beyond his control." The International Covenant on Economic, Social and Cultural Rights (1966) art. 11 is much more explicit "1. The States Parties to the present Covenant recognize the right of everyone to an adequate standard of living for himself and his family, including adequate food, clothing and housing, and to the continuous improvement of living conditions. The States Parties will take appropriate steps to ensure the realization of this right, recognizing to this effect the essential importance of international co-operation based on free consent. 2. The States Parties to the present Covenant, recognizing the fundamental right of everyone to be free from hunger, shall take, individually and through international co-operation, the measures, including specific programmes, which are needed: (a) To improve methods of production, conservation and distribution of food by making full use of technical and scientific knowledge, by disseminating knowledge of the principles of nutrition and by developing or reforming agrarian systems in such a way as to achieve the most efficient development and utilization of natural resources; (b) Taking into account the problems of both food-importing and food-exporting countries, to ensure an equitable distribution of world food supplies in relation to need".

[2]The International Covenant on Economic, Social and Cultural Rights (1966) art. 15.1 (c) "To benefit from the protection of the moral and material interests resulting from any scientific, literary or artistic production of which he is the author" on the protection of economic interest. Here the UN General Comment on this section specifies that the other affected rights of the Covenant and "article 15, paragraph 1 (c), is at the same time mutually reinforcing and reciprocally limitative" (UN Committee on Economic Social and Cultural Rights 2006, par. 5).

and the availability of technological solutions for those with the greatest hunger and malnutrition burden (De Schutter 2009b). In the last two decades, we have observed the emergence of social justice arguments in relation to scientific malpractice and biopiracy, calling for informed consent and fair access and benefit-sharing procedures (De Jonge 2009; Wynberg et al. 2009), and measures to limit the fabrication and falsification of data (Resnik 2009). In this book, I will claim that this analysis is too narrow, as it only focuses on issues of justice in exchange and distributive justice, with some aspects of procedural justice, and that in order to develop truly socially and environmentally sustainable agricultural policies, a wider normative framework is needed. Science policies also affect participation in scientific projects and their governance, the availability of resources for future generations, and how injustices are handled in the technology development processes, during the technology's introduction and after its use. The aim of this book is to develop such a wider normative framework by examining the hypothesis that agricultural innovation and governance faces a six-dimensional social justice problem, as it involves six categories of justice:

(1) *Justice in exchange*: dealing with issues of access to the objects of innovation and fair rewards for innovators.
(2) *Distributive justice*: securing a fair share of research attention in proportion to the urgency of people's needs.
(3) *Contributive justice*: defending an inclusive research environment that encourages and facilitates widespread participation.
(4) *Intergenerational justice*: making sure that current generations provide adequate public goods to compensate for their consumption of exhaustible or slowly regenerating resources relevant for food production.
(5) *Procedural justice*: establishing rules and process guidelines to improve the impact of innovations and avoid negative outcomes.
(6) *Restorative justice*: ensuring that any injustices that emerge during scientific and technological innovation processes are adequately dealt with.

These six dimensions, applied to the wider set of problems and opportunities we encounter in agriculture, single out the various demands of what I will refer to here as "agricultural justice". The first four forms of justice—justice in exchange, distributive justice, contributive justice and intergenerational justice—principally have intrinsic value, that is, they should be pursued for their own sake, as a matter of justice. The last two dimensions of justice—procedural justice and restorative justice—have both and instrumental and non-instrumental value. They serve to make sure that the demands of the first four dimensions are met, but are also valued on their own, as they build social cohesion, provide room for participation, and allow people to contribute to change. Before going into any more details of this normative framework, let me introduce and contextualize the problem field.

To begin, we must fully acknowledge that many of the most problematic negative consequences of establishing the present intellectual property regimes as the main formal innovation incentive systems are due to extreme inequality. We live in a world of extreme inequalities, be they in purchasing capacity (Milanovic 2011; Piketty 2014), availability of genetic resources (Kloppenburg 2005), scientific infrastructure and capabilities (Timmermann 2014b; Mazloumian et al. 2013), or social stability and institutional strength (Pogge 2008). Extreme inequality is a continually contributing factor that hinders any full realization of the six dimensions of social justice.

The innovation incentive systems that are in place directly shape the types of innovation that will or will not be available, and how science is done. They have a direct effect on people's lives in terms of the means of production available, the future course of research, the effects of technology on people, their relationships to others and the environment, and the range of technological solutions that are available for adapting to current and upcoming challenges. With such a profound impact on people's lives, in particular in relation to something as vital and indispensable as food production, scientific and technological development needs to become more democratic. This can be achieved by opening up its decision-making mechanisms and being more inclusive in its development processes, so as to be more socially sustainable and increase people's willingness to cooperate by sharing data, samples, feedback and observations. If people perceive a system as unjust, then they will deem it unworthy of their cooperation (Ooms 2010; Ruger 2015).

Agricultural innovation is far from being perceived as a just global cooperative system. Our worldwide intellectual property regimes give a commercial advantage to the type of innovation propagated by the Green Revolution, as these meet the formal patent requirements: (i) an inventive step, (ii) novelty and (iii) industrial applicability, or are sufficiently homogeneous and stable to qualify for plant breeders' rights (Louwaars et al. 2013). Intellectual property regimes take their current form primarily to satisfy the needs of the industry of the Global North, and are nowadays globally enforced through the World Trade Organization's (WTO) trade dispute settlement mechanisms and bilateral trade agreements (Drahos and Braithwaite 2003; James 2012). As a consequence, intellectual property creates a strong incentive to concentrate on delivering technological solutions that can be sold to consumers with high purchasing power, while neglecting financially weaker markets regardless of their size in terms of the number of people (De Schutter 2009b). It also creates incentives to sell more tokens of a lucrative crop variety than are ecologically sustainable (Timmermann and Robaey 2016). Such large monocultures are more vulnerable to pests and are less resilient to extreme climatic events such as hurricanes (Holt-Giménez 2002). When prices rapidly increase due to crop losses, it is usually the people with the lowest disposable income that pay the price of inadequate agricultural policies with their hunger (Ziegler 2011).

Despite these problems, we should not ignore the fact that proprietary science has also delivered a number of technological solutions for growing certain food crops in places where this was not possible before, as well as allowing the extensive automation of greenhouses and developing plant varieties for bioremediation. Improvements in the automation of food production systems have reduced reliance on human labour, which has accordingly been freed for use in other productive enterprises whenever these were available (Mazoyer and Roudart 2006). It has also increased yields so substantially that agricultural land can now be given back to nature, allowing the capture of carbon and the harbouring of wildlife (Balmford et al. 2018). An increasing technologization of agriculture, through digital farming, could considerably lower the current environmental footprint of conventional agriculture, by reducing chemical spills, using resources more efficiently and detecting pathogens at an earlier stage (Bongiovanni and Lowenberg-DeBoer 2004; Gebbers and Adamchuk 2010). While acknowledging the benefits of proprietary science, we need to discuss policies that would remediate its shortcomings and promote its benefits, and a normative framework that supports and evaluates such measures.

Over-reliance on intellectual property incentives leads to a situation where a variety of innovations that are socially and environmentally desirable are currently produced at an insufficient rate or not developed at all (Vanloqueren and Baret 2009). Intellectual property regimes, as currently conceived, generally do not facilitate the protection of heterogeneous and incremental innovation. New plant varieties that are not stable and show a degree of variation do not qualify for protection, reducing their commercial uptake and thus negatively affecting agrobiodiversity (Halpert and Chappell 2017). Moreover, many methods of agroecological farming are also not patentable, as these are based on principles rather than inventions. This makes it difficult to incentivize and propagate agroecological innovations and the maintenance of agrobiodiversity without adequate programmes from governments, NGOs or international organizations (Tittonell 2013; Nicholls and Altieri 2018). When nobody can secure exclusivity for an invention, it becomes much more difficult to recoup the costs of diffusing innovations, with the unfortunate result that certain technologies and techniques end up being underused even when they are freely available. However, this does not mean that we have to expand the scope of what is patentable. Non-proprietary science can be supported through prizes and government funding (Love and Hubbard 2007; Hollis and Pogge 2008; Timmermann 2014a; Outterson 2014), thus stimulating parallel research streams and approaches that are much more receptive to the classical Mertonian scientific norms: disinterestedness, organized scepticism, communalism and universalism (Koepsell 2010; van den Belt 2010). This would allow the participation of people who oppose ownership of genetic resources as a matter of principle, as well as those who are unwilling to cooperate in a research environment that is not open and inclusive. Nevertheless, we should maintain the same level of scrutiny when assessing technologies that were developed without intellectual property incentives, as they are not free from social justice problems. While financial incentives may tempt businesses to market unsafe products, companies who have worked hard to gain a positive reputation have also an incentive to avoid jeopardizing their brand name by selling hazardous or ineffective products. In contrast, the absence of technology ownership may create a situation where nobody assumes responsibility.

It has long been acknowledged that inventing is not a morally neutral act (Godin 2015). Our global capacity to innovate can avoid the occurrence of problems, mitigate their negative effects, and also create new forms of harms. Perhaps the most established approach in technology assessment is to focus on eventual harms to people, nature and social relations. Harm may consist in jeopardizing the existence of, inflicting suffering on, or even altering the integrity of living beings, cultural objects and natural environments. Moreover, innovation may even change the moral character of a harmful event (COMEST 2015). By making a technological solution available, a specific harm can be avoided, provided that one has sufficient resources and liberty to acquire or duplicate the invention. Here we need to be aware that suffering an avoidable harm is normatively and emotionally different from suffering an unavoidable harm: unequal access may have deep effects on social interactions. Due to its vital importance to people's lives, agricultural innovation has been subject to an enormous amount of scholarly criticism and reflection. The social studies of agricultural innovation have strongly benefited from the technology assessment literature. I therefore provide a short overview of the most influential discourses, to highlight the strength of using a six-dimensional social justice perspective.

1.2 The Contemporary Roots of Philosophically-Inspired Technology Assessment: Ethical, Legal and Societal Aspects, Responsible Research and Innovation, and Sub-disciplinary Approaches

There have been a number of schools of thought that have worked on technology assessment. From early on, people condemned the effect that industrialization was having on the way we relate to natural landscapes, with perhaps Aldo Leopold's "A Sand County Almanac" (1949) being the most influential example. In the aftermath of the catastrophic events of the two world wars, the idea that technology could annihilate humankind as a whole started to become a major concern for rich and poor alike. Early technology assessment was particularly worried about the harms that technology could bring to humanity, social relations and the environment (Jonas 1979). The threat of nuclear war and disasters acted as a major discussion changer (Grunwald 2009), as humans had to clearly accept for the first time that they were capable of threatening their own very existence as well as that of much life on Earth (Beck 1986). The deeply anchored belief that the Earth was too mighty to be damaged by humans had to be given up. Particularly in the life sciences, there was an early concern about which interventions could be considered natural and which not, triggering a discussion on the very question of what is to be considered part of nature (Kaebnick 2007; Siipi 2008; Van Haperen et al. 2012). The interest in these types of questions and the urgency of their study grew so extensively, that it led to the establishment of philosophy of technology as a special field of study in the 1960s and to science and technology studies (STS) as a separate discipline in the 1980s (Franssen et al. 2018).

In the late 20th century, technology assessment expanded its scope and methods of analysis considerably, especially as the use of modern technologies became much more invasive and extensive (van Est et al. 2012). The study of Ethical, Legal and Societal Aspects (ELSA)—or Ethical, Legal and Social Implications (ELSI)—of new technologies broadened its scope to assess how technology could affect people, society and nature, and be used or adapted to meet specific social goals. An inter-disciplinary dialogue and analysis was pursued with the goal of finding out the effects, potentials and risks that technologies would have for society and nature (Zwart et al. 2014). This dialogue encouraged intensive interaction between scholars from the natural and social sciences, often demanding extensive empirical studies of innovation sites.

After the turn of the millennium, a new perspective on social technology assessment started to gain increasing momentum: Responsible Research and Innovation (RRI) (Rip 2014; Schomberg 2015; Owen et al. 2012). To assess whether an innovation process was conducted responsibly, four dimensions need be evaluated: anticipation, reflexivity, inclusion and responsiveness (Stilgoe et al. 2013). Others have also argued for adding the dimension of transparency (Ravn et al. 2015). Here "anticipation" means that innovators need to identify and minimize unintended consequences of their innovations, "inclusion" seeks to incorporate a wider set of stakeholders into deliberation processes while at the same time involving them more intensively in innovation assessment, "reflexivity" requires being critical of one's own motivations and assumptions while also considering the perspectives of others, and "responsiveness" refers to the ability to adapt and change innovation processes according to societal needs and new knowledge (Eastwood et al. 2019). Meanwhile, "transparency" requires access to results and information on research processes in order for societal actors to be able to engage in a critical assessment and prepare their innovation advocacy or objections. The idea is for innovators and their funders to engage with societal actors in an interactive assessment process regarding their technologies and their effect on society and the environment (Schomberg 2012). Here the innovator becomes the central actor in making sure that technologies gain public acceptance (Robaey and Timmermann 2018). RRI is not just a complex procedure to be followed as a matter of corporate social responsibility; it may also be pursued for purely instrumental reasons, in order for technologies to be better received, accepted and diffused through society, as social opposition to the technology will be handled at a much earlier stage (Asveld et al. 2015). Although the idea of RRI comes from a European context, it is also slowly being adopted in other parts of the world (Macnaghten et al. 2014).

In parallel to this scholarship, there are three areas of philosophy that deal with questions of innovation and society outside philosophy of technology: philosophy of science, ethics, and political philosophy.

Philosophers of science have been particularly keen to show how certain structures and incentive systems favour the development of one scientific approach or method over another. Perhaps the strongest complaint within this tradition is to accuse conventional farming of reductionism. In agriculture, this can be seen in failing to take wider (and also smaller) elements of a system into consideration, in particular overlooking the importance of soils in agroecosystems (Thompson 2017). Philosophers of science also criticise the emphasis on technoscience, or forefront science, to develop the innovative solutions the world really needs, instead of concentrating on preventive measures or more resilient designs (Reiss 2010; De Winter 2012). For example, the choice of developing genetically modified crops was most likely not the fastest way to improve food security for the poor (Lacey 2012). Another point of criticism that philosophers of science have raised concerns the distorting effect that the commercialization of science is having, by impeding a democratic setting of research agendas, independent assessment and even how research results are presented (Fernández Pinto 2015). When technology assessment or biosafety controls raise epistemic concerns, they can have direct consequences for people and the environment when technologies are released or favoured on the basis of incorrect information on safety and efficacy (Reiss 2010). Since the distortion of science has effects on social justice, this book will discuss relevant considerations made by philosophers of science.

Another philosophical area that contributes to science and technology assessment is ethics, in particular the perspectives that are strongly influenced by biomedical ethics. Contributions that fall into the broader area of scientific integrity and research ethics are especially concerned with issues of scientific malpractice, such as failures to obtain informed consent, deceit, plagiarism, and data fabrication and falsification (Koepsell 2016; Resnik 2009). This analysis relies on principles with a strong tradition in bioethics that innovators should adopt, such as assuming responsibility, being cautious, respecting people's autonomy and sharing the benefits of research (Linares Salgado 2013, Lolas 2006). Strong ethical codes, regulations and review committees are being established to hinder and punish scientific malpractice. These findings are a major contribution to identifying issues of social justice.

Nowadays, we are also witnessing the consolidation of additional approaches in the study of the social impact of innovation, by working with theories of justice. These theories can highlight the social justice issues that innovation raises by incorporating new insights from political philosophy and theory. Users of such theories fall into two main categories: philosophers and political theorists who analyse contemporary social justice issues regarding the use of intellectual property rights, and the interdisciplinary groups who work on environmental justice and those inspired by this discourse.

Political philosophers have criticized intellectual property regimes from different angles. Their analysis concentrates both on topical issues and on the application of specific principles (Gosseries et al. 2008; Lever 2012). For example, those inspired by John Rawls have questioned the legitimacy of intellectual property regimes, as these create inequalities that do not improve the situation of the worst off (Dumitru 2008). Following this line of thought, it has also been argued that innovation not only fails to improve the lives of the worst off, but is also a missed opportunity to satisfy their basic rights. Indeed, innovations may bring inequalities that cause further destabilizing factors for poverty reduction and democracy (Buchanan et al. 2011; Papaioannou 2018). A harder line of criticism is raised by Thomas Pogge (2008), accusing the institutions that have established intellectual property regimes around the world of violating the negative duty of not harming the poor, as these institutions deprive the global poor of access to vital innovations and divert research attention from their urgent needs towards the provision of goods for more lucrative markets. Similarly, Nicole Hassoun (2016) links issues of structural injustice to the need to inform consumers, so that they can exercise responsible consumption and thereby encourage corporate social responsibility. Others have used the capabilities approach to condemn intellectual property regimes that insufficiently contribute to facilitating the provision of resources required for securing basic needs, such as food, and are not sufficiently inclusive to allow people to meaningfully participate in projects that concern them (Clague 2006; Sunder 2012). Inspired by recognition theorists, such as Nancy Fraser, others condemn scientific enterprises that fail to recognize the inventiveness of some actors, particularly small farmers, while over-recognizing the work done by others, mostly biotechnologists and commercial breeders (Coolsaet 2016; Timmermann 2013). Another approach, followed by those influenced by the work of Iris Young, is to highlight the effect of structural injustices, such as the ability of stronger parties to shift a discourse in particular direction, as in the case of concentrating the debate on GM foodstuffs on issues of safety, leaving for a second stage concerns about power and dependency in food regimes (Hicks 2017). These lines of argument have been critically examined, and many of them will be further developed throughout this book, in relation to the respective categories they fall into.

Another discourse with a strong background in political philosophy and theory is represented in the debate on environmental justice and the newer fields it has inspired: energy justice, climate justice and food justice. Since these discourses deeply incorporate social justice language, it is worth looking at them in more detail.

Environmental justice was born out of a strong social repulsion for how environmental burdens were distributed among society. In the case of the United States, research showed that African American communities in particular were overwhelmingly burdened with environmental harms, particularly the disposal of toxic waste and chemically hazardous facilities, while receiving little benefit from these arrangements (Shrader-Frechette 2002). In order to provide a normative framework for opposing such outcomes, scholars working on environmental justice have made use of the concepts of distributive justice to argue for a fairer distribution of burdens and benefits, and of the concepts of participatory justice to argue for a wider inclusion of people from local communities when deciding on the allocation of environmental burdens and benefits (Figueroa and Mills 2001). This discourse has been expanded over the last two decades, so as to acknowledge the importance of individual and community capabilities to empower people to protect the natural environments and heritage sites that they are customarily bonded with (Schlosberg and Carruthers 2010).

A more recent perspective comes from scholarly work on energy justice. The assessment offered here distinguishes between distributive justice, recognition justice and procedural justice (Jenkins et al. 2016; Pesch et al. 2017). As with environmental justice, by using ideas of distributive justice, energy justice seeks a fairer distribution of the ills of electricity production and a redistribution of its benefits. Recognition justice is used to highlight the main problems of misrecognition, particularly the problem of not involving local communities in decision-making and not recognizing people from such communities as possible collaboration partners or potential service providers. Procedural justice seeks to establish a decision-making and problem-solving mechanism that fairly includes the affected parties.

Another emerging field is climate justice. While also strongly inspired by the environmental justice literature (Jenkins 2018), the nature of climate change has obliged this discourse to critically engage with issues of global justice and historical contributions (Patterson et al. 2018). This has led scholars working in this area to build on other forms of justice, such as intergenerational justice and corrective justice.

Thematically, the discourse that is closest to the object of the present study is food justice. The field of food justice has grown significantly in the United States over the last two decades, due to increased awareness of the fact that the price of bad food politics is overwhelmingly paid by racial minorities and the poor. This has led the discourse to concentrate around issues of injustice, particularly the unfair disadvantages that are suffered due to race, gender, class and youth discrimination in all aspects of the food system—production, distribution, preparation and consumption (Glennie and Alkon 2018).

Agricultural innovation involves highly dynamic processes that affect multiple actors of highly differential sizes and strengths in interaction with people all over the world. As such, this book will build on these different approaches to construct a new assessment tool for a broader set of dimensions of social justice. When applying the justice discourse to agriculture, we are confronted with the question: what does "agricultural justice" demand?

1.3 Innovation, Agricultural Justice, and Its Six Dimensions of Social Justice

Due to the intellectual history of this concept, when I claim that we need a broader assessment of potential social justice issues, I owe readers an explanation. Why six dimensions, and not more, or less?

The observation that agriculture involves a series of social justice problems is of course old. The concept of agrarian justice has been around for quite some time. Condemning as unjust the extreme poverty that results from land enclosure has been prominent since at least Thomas Paine's pamphlet "Agrarian justice" (1797). In today's terms, Paine saw in the highly unequal allocation of land ownership at the moment of one's birth an issue of distributive injustice that needed to be remediated (ibid.). The same concept of agrarian justice has more recently been used to illustrate the role of law in promoting social justice, such as the right to food to specify food entitlement, anti-discrimination laws to promote equal legal access to land and credits, food safety laws, and environmental safety and labour protection laws (Hamilton 2008). Another research line uses the related term "land justice" to call attention to issues of land reform, labour rights, environmental sustainability and climate change adaptation (Williams and Holt-Giménez 2017).

Due to the subject of this book, I have departed from these earlier studies and incorporated into my proposed framework the principal perspectives of the social justice studies of innovation, particularly insights from the Responsible Research and Innovation literature. The central objective of addressing the demands of justice in agricultural innovation is to secure cooperation and facilitate the alignment of research agendas towards developing socially necessary and environmentally sustainable goods. This obliges us to depart from traditions that are only willing to see the state as an addressee of justice, and instead treat perceived injustices as outcomes of both the state's and private parties' actions and omissions. Taking this broader approach, we must consider which individual, group or state actions and omissions influence, individually or in aggregate, the extent to which innovation systems are perceived as just. It is crucial for social and environmental sustainability that people cooperate with agricultural innovation systems, but they will be reluctant to do so if these systems are perceived as unjust. Identifying the injustices involved in such systems is the first step towards addressing them.

Let us briefly review the highlights of the literature on innovation systems assessment that uses the language of social justice, in order to contextualize the additions and modifications that I will suggest in this book to justify the introduced six dimensions of social justice. In the following section I will distinguish four forms of justice: justice in exchange, distributive justice, contributive justice and intergenerational justice, as well as two approaches to implementing these demands and redressing failures to realize them: procedural justice and restorative justice.

As briefly mentioned before, the discussion of issues related to innovation and social justice has concentrated on the so-called availability and accessibility problems. These two problems were the most pressing issues in the access to medicines campaigns (Hollis and Pogge 2008): we need fairer prices to make sure that people can have access to medicines, and we need to distribute research attention so that medicines for treating the urgent health needs of the poor become available. Many have characterized access to medicine as an issue of distributive justice, as it deals with the allocation of medicines for the poor. There are, however, good arguments for analysing this problem of access as an issue of justice in exchange in our current world order. As will be discussed in much more detail later on (see Chap. 2), one of the key arguments to justify intellectual property rights is that they may lead to just rewards for innovation, thus positing a strong link between exclusive rights and deserved rewards. As long as innovations are treated as the fruits of people's labour, and not as common property or as something that society has a legitimate claim on, the issue of proportionality and fair rewards is crucial, as disproportional rewards are incompatible with universal access. Moreover, distributive justice usually deals with goods that already exist (Schroeder and Pisupati 2010). In order to distribute a good, it first needs to be available. However, as access to medicine activists have clearly identified, there are still a number of deadly diseases that have not received proper research attention, and for which it is still the case that no proper treatment is available (Drugs for Neglected Diseases Working Group 2001). Agriculture suffers a similar fate, as many key tropical food crops have not received sufficient attention from plant breeders. Similarly, tools and machines for facilitating the work of low-external-input farming systems need to be developed and improved. Moreover, innovations specially tailored for resource-poor settings are needed (Bragdon 2016). Interpreting distributive justice as demanding a fair distribution of benefits implies that poor people around the world should benefit from scientific advancement as well.

Distributive justice demands a stronger reorientation of research attention towards the urgent needs of the poor.

A new perspective I propose for this type of analysis is to strongly emphasize the issue of participation by working with the concept of contributive justice. The idea behind contributive justice is to allow people to develop their capacities and contribute to society according to their abilities (Timmermann 2018). The central goal of this dimension of social justice is to allow wider and more intensive participation and thereby empower people to contribute towards improving social well-being (Britz and Lipinski 2001; Gomberg 2007). The demand to raise participation to the status of one of the basic rights is not alien to political philosophy. Henry Shue (1996), to name a well-known example, claimed that participation is fundamental to the enjoyment of the other basic rights, such as access to subsistence goods, as it allows citizens to monitor whether political institutions and private parties are contributing effectively to their continuous fulfilment. Also, in the debate on intellectual property rights, there are voices that have given participation a central role. Matthew DeCamp (2007), for example, has condemned the extremely uneven concentration of intellectual property rights, which allows established companies to block the eventual competition of newcomers by engaging in practices that hinder follow-up innovation, such as demanding usurious or unrealistic fees, or simply not licencing certain technologies. Using the concept of contributive justice, I back a key principle of Responsible Research and Innovation—inclusion—and a central worry of bioethicists who seek to protect autonomy through empowerment. Demanding and encouraging participation is also crucial for recognition justice—as stipulated, for example, in the energy justice discourse. Meaningful participation in joint projects allows people from different sectors to see each other as peers (Fraser 1998). Wider participation in the sense defended here also counters one of the main worries of distributive justice theorists: society needs to produce goods (i.e. innovate) before it is able to distribute them. Increasing scientific participation under conditions of collaboration leads to improved innovation capacities that can address a wider array of social problems (Torrance and von Hippel 2015).

Another form of social justice I argue for that needs to be taken into systematic consideration is intergenerational justice. The size the human population has reached has made it mandatory to systematically assess the pressing issues of social and environmental sustainability as issues of social justice. In a world of extreme inequality, environmental footprints are not absorbed evenly by the world's population: the poor are those who are most likely to live in polluted areas and suffer the most when food production collapses. As this is a foreseeable outcome, any innovation system must take issues of long-term sustainability into consideration: an issue strongly embraced by Responsible Research and Innovation scholars in their rubric of "anticipation" and "responsiveness". In addition, we have reached a degree of environmental impact that is threatening our ability to survive as a species, as the case of climate change and its disastrous effect on food systems shows (McMichael 2017). Principles of reciprocity also demand that we offer future generations similar opportunities to live a flourishing life: something that sets limits on how we use resources and how we affect the environment, and also imposes duties to compensate future generations for the resources we have consumed (Beekman 2004). The urgency of addressing environmental issues in agriculture has made it necessary to treat issues of sustainability as a separate category, as they go beyond issues of intergenerational distributive justice. We are not only destroying irreplaceable goods, but also jeopardizing ways of life whereby living in non-parasitic relations with nature is central for communal self-understanding.

To ensure that these four forms of justice are realized, I propose the systematic inclusion of two further dimensions of justice: procedural justice and restorative justice.

Procedural justice aims to establish a series of guidelines and processes for ensuring fair outcomes. In applying this concept to our field of analysis, we must acknowledge that agriculture involves a very large number of actors with highly unequal bargaining power, and that some groups (particularly racial and ethnic minorities, indigenous people and women) are much more vulnerable due to their historical discrimination and sometimes even inequality before the law. To avoid injustices, we need a series of guidelines for avoiding scientific malpractice, exploitation and the development of socially and environmentally unacceptable innovations, and secure sufficient space for self-determination and participation. This dimension analyses many of the key points that Responsible Research and Innovation addresses under the categories "anticipation", "reflexivity" and "responsiveness".

The last dimension of social justice I argue for, to make sure innovation has a socially and ecologically desirable effect, is restorative justice. Science is a social enterprise and collaboration is increasingly the norm (Jefferson 2006). There is increasing doubt concerning whether individual scholars deserve the full credit for their contribution to science, as the label the "myth of the sole inventor" suggests (Lemley 2012). This even applies to historical inventors as important as Thomas Edison, who pioneered an "invention factory", hiring a large number of scientists to work on diverse projects (Drahos and Braithwaite 2001). Science historians continuously reveal the role of ignored and underrecognized scholars, in particular the contributions of female and non-white scientists.

Recently people who assess innovations through a Responsible Research and Innovation lens, or from a bioethics perspective, have opted to focus on preventing injustices by carefully drafting procedures for avoiding bad scientific practices, securing informed consent and preventing technologies that have a harmful effect on society and the environment. These measures are, however, hard to apply to incremental innovation or, more generally, to innovation processes that incorporate ideas, observations and worries over long periods of time and extended regions in informal settings, as seen with many innovations developed by smallholders. This makes restorative measures especially suitable for injustices committed within larger groups, thus they are a necessary complement for establishing justice.

As is the case with any group work, it is foreseeable that injustices of various degrees and dimensions will occur that require reconciliation measures: either due to conscious wrong acts, unawareness, naivety or negligence, or even just disagreement or misunderstandings. Injustices demand restoration measures, and the absence of an arbitration system or processes is an injustice in itself.

This book applies these forms of justice across national borders. A cosmopolitan perspective is defended on three major aspects: human rights, interdependence, and the importance of securing cooperation. According to the traditional human rights perspective taken from the Universal Declaration of Human Rights, securing the human right to food is primarily the task of the state where people reside, and is complemented by all nations when the demands of that right are not met. This creates obligations for all states, and arguably also individuals, to assist. The second argument is one based on the interdependence of international agriculture and genetic flows. We cannot separate the different contributions that each country has made to agricultural knowledge, nor can we accurately measure how much each nation has done to protect varieties (Mazoyer and Roudart 2006). This has created a strong interdependence and made agriculture, for political considerations, a type of common international cooperative "institution" upon which we all rely and arguably also have certain obligations towards. Strictly speaking, no country could legitimately claim that "they properly deserve the full marginal benefit of their superior social productivity" (Hinsch 2001, 74), when they rely (and have relied) on the free work of smallholders to conserve such a vital resource as agrobiodiversity and count on ecosystem services originating in other nations (IPBES 2019). The third point concerns the need to secure cooperation. Here it is not so important whether institutions are legally or philosophically considered just, but rather whether the general public can reasonably consider them to be just. If people do not consider agricultural research institutes to be worthy of cooperation, then this may create severe problems when seeking samples, observations, or assistance in halting the propagation of pathogens.

The framework proposed here aims to systematically assess technologies in the area of agriculture by focusing on theories of justice. This book offers an examination of the extensiveness of this framework. Before moving on, some key concepts need to be clarified.

1.4 On Science, Technology, Invention, Innovation, and Other Key Concepts

As is customary, the main concepts of this book deserve a proper introduction. Traditionally, "science" refers to the theoretical understanding of nature, and "technology" to the application of this knowledge. Scientific advancement in the area of the life sciences has blurred this distinction, creating a significant overlap between science and technology (Eisenberg and Nelson 2002). Agriculture is no exception to this development. On the contrary, agroecology, for example, has been referred to as a "science", a "practice" and a "movement", thus refusing to keep science and its applications apart (Wezel et al. 2009). The farmer becomes an innovator through on-farm experimentation, and by exchanging observations and knowledge with peers and researchers. The combination of basic and applied research has a strong tradition and has been referred to as "Pasteur's Quadrant", honouring Louis Pasteur's work in use-inspired basic research (Doran et al. 2017). This combination of research and development has an impact on who does research and how research is done.

As further implications, we can observe a hesitance to consider certain methods and approaches as "scientific", as well as a hierarchy of what is to be considered "more scientific" science (Tierney and Holley 2008). This problem is also partly due to the fact that the various sciences are much more different than previously thought. Attempts to unify science by appealing to a unique method or the like have been discarded (ICSU Study Group on Science and Traditional Knowledge 2002). A very broad but, for the purposes of this book, very enlightening understanding of science is Paul Hoyningen-Huene's (2008) description of science as pursuing and aiming at systematicity. Science characteristically seeks a full understanding of a subject matter, by systematically elaborating descriptions, explanations, predictions, a defence of knowledge claims, epistemic connectedness, and the representation of knowledge, while seeking an ideal of completeness, in relation to what the subject allows (ibid.). Thus we can understand farmers' observations as scientific when they are systematic and seek vast understanding. In contrast, observations that are spontaneous, unmethodical and intended only to satisfy momentary curiosity do not count as a scientific.

The concept of technology is very broad in its scope, and is traditionally practice-oriented, as its Greek root, *tekne* (art, craft), reveals. "Technology" refers to something that was created or can be used for a purpose, has a function and leads to benefits. In addition, it is often, but not always, human-made; even though it can be made from naturally occurring components, required skills or art for its creation, has a physical form, and is the result of primary or secondary intentions (Carroll 2017). Throughout this book I use "technology" or "object of innovation" in this very broad sense, irrespectively of whether something originates in a research laboratory or a farm.

As its title indicates, the concept of innovation is central to this book. "Innovation" is also a very broad concept and only in recent decades has it been strongly linked to technological innovation (Godin 2015). In this sense, innovation is often understood as a technological invention that is used and adopted, regardless of whether it is sold, exchanged or given away freely (Gault 2012), in contrast with the countless inventions that never leave their creators' shelves. Strictly speaking, there are innovations in all fields of knowledge, from linguistic innovations to legal innovations, and from farmers' innovations to innovations in the exploitation of nuclear energy. Here I use "agricultural innovation" to refer to improvements and methods that have the aim of making more food available, either by improving the quality and quantity of yields, or by avoiding food waste and destruction. Non-food-related agricultural innovations will be only mentioned in passing. While I focus on innovations of a technical and technological nature, some social innovations will also be discussed, such as the interesting legal innovations that have been developed to protect the intellectual commons through creative commons licenses, and innovations that support the diffusion of food production improvement methods.

The subject of this book is agricultural innovation, and I make repeated reference to five food production streams: conventional agriculture, digital farming, organic agriculture, subsistence agriculture, and agroecology. Under "conventional agriculture" I categorize farming systems characterized by the use of improved seed varieties, heavy machinery, synthetic pesticides and fertilizers, and monocultures; in other words, those innovations that are commonly associated with the Green Revolution, and feature high external input systems. In contrast, when I speak of "digital farming", I refer to production systems that have incorporated many of the innovations of conventional farming, but also make use of information technologies to minimize the destruction of non-target organisms and the spill-offs of agricultural inputs. This type of agriculture is often labelled "precision agriculture", but I propose to avoid using that term, as it gives the impression that this is the only precise farming method. When I speak of "organic agriculture" I mostly include those farming systems that have substituted synthetic inputs with those that are accepted by organic certifiers, without considering a redesign of the production system (Rosset and Altieri 1997). By "subsistence agriculture", I mean the type of agriculture that does not use external inputs or improved seed varieties, and does not significantly take advantage of ecological interactions, mostly due to poverty, lack of training, and insufficient knowledge exchange. I use "agroecology" to refer to farming systems that make efficient use of ecological interactions, develop their own inputs, and aim to close nutrient cycles, hence aiming for low-external-input production schemes (Altieri 1989). Lastly, this book concentrates on agriculture for food and feed production, as these have stronger implications for social justice due to the human right to adequate food. I will nevertheless make occasional references to agriculture directed towards the production of alcoholic beverages, medicines, biofuels and manufacturing materials.

The choice of innovation stream is not the only relevant issue for social justice; also relevant is how innovations are made and diffused. Depending on how innovation processes are designed and understood by their participants, they can include farmers in or exclude them from knowledge acquisition and technology optimization processes (Klerkx et al. 2012). The same applies to technology diffusion processes. If technology uptake is seen as a part of the innovation process, so that technology developers work with farmers to optimize technologies and learn about user innovation, the process becomes much more inclusive and better suited to the real world (Röling 2009). Understanding inclusive technology development as an ideal of social justice is particularly important for innovation streams that do not rely on knowledge-intensive farming, since they rarely integrate farmers systematically into technology development and improvement. Considering these factors, I will deal with innovations broadly, including their choice, process and diffusion.

There are two other terms I need to justify: "the Global South" and "the Global North". There is a long-standing debate on how countries are to be grouped depending on their level of industrialization, realization of human rights, institutional strength, economic freedom, market size and rule of law, among other factors. While this division is far from satisfactory and ignores the huge range of differences within countries, in this book I will say "the Global South" to refer to those countries that were traditionally labelled as "Third World Countries" or "the Developing World".

I recognize that there are major differences between the countries that fall into this category, and worse, that by grouping countries in this binary way I may imply a valorisation of "good" and "bad", or even that there is a single form of social organization that is universally desirable (Figueroa and Mills 2001). These labels also imply that all countries are moving in a desirable direction, which unfortunately is also not correct (Rodney 1972). There is also the problem that some countries are officially included in one category, mostly because of their gross domestic product, but have failed to establish or maintain the institutions that are important for my analysis, such as scientific infrastructure and freedom of speech. Nevertheless, I will use this distinction to indicate countries that meet a certain threshold of realizing human rights and acknowledge, as far as possible, that economic wealth is not an accurate indicator for revealing which countries fulfil their human rights commitments, even more so when we are aware of the wide range of protections that human rights involve, such as freedom from hunger, political rights, and participation in science, and many other rights affirmed in the Universal Declaration of Human Rights (1948), the International Covenant on Economic, Social and Cultural Rights (1966) and the International Covenant on Civil and Political Rights (1966). Therefore, the characteristics I attribute to the countries of the Global South amount to not meeting a certain threshold in the following categories: (i) financial capital, (ii) scientific infrastructure, (iii) rule of law, (iv) strong democratic institutions, (v) free speech and access to information, and (vi) freedom of movement. These traits have a substantial effect on agricultural innovation systems, and are therefore worthy of special attention. Lastly, when I speak of "the poor", I mean financially poor, unless otherwise specified.

Also, when I say "we", I engage with the reader as an agent capable of contributing to change, even if only in a minor way, irrespective of background(s).

1.5 Overview of the Book

In the following pages, I dedicate two chapters to the effects that intellectual property or the lack thereof have on agricultural innovation. The first of these chapters deals with the different justifications for intellectual property and its different regimes (Chap. 2). The second one discusses the advantages and disadvantages of innovation systems that function without exclusive rights, and describes two such systems: open science and traditional knowledge systems (Chap. 3). Readers familiar with the innovation and intellectual property debate may choose to skip these two chapters, as they will already understand the background context of the next chapters. Thereafter, I present each of the different social justice dimensions and its effects on agricultural innovation in its own respective chapter: justice in exchange (Chap. 4), distributive justice (Chap. 5), contributive justice (Chap. 6), intergenerational justice (Chap. 7), procedural justice (Chap. 8), and restorative justice (Chap. 9). Lastly, in Chap. 10, I highlight the advantages of having parallel innovation systems for securing the demands of social justice.

References

Altieri, Miguel A. 1989. Agroecology: A new research and development paradigm for world agriculture. *Agriculture, Ecosystems & Environment* 27 (1): 37–46.

Altieri, Miguel A., Clara I. Nicholls, and Rene Montalba. 2017. Technological approaches to sustainable agriculture at a crossroads: An agroecological perspective. *Sustainability* 9 (3): 349.

Asveld, Lotte, Jurgen Ganzevles, and Patricia Osseweijer. 2015. Trustworthiness and responsible research and innovation: The case of the bio-economy. *Journal of Agricultural and Environmental Ethics* 28 (3): 571–588.

Balmford, Andrew, Tatsuya Amano, Harriet Bartlett, Dave Chadwick, Adrian Collins, David Edwards, Rob Field, Philip Garnsworthy, Rhys Green, and Pete Smith. 2018. The environmental costs and benefits of high-yield farming. *Nature Sustainability* 1 (9): 477–485.

Beck, Ulrich. 1986. *Risikogesellschaft: Auf dem Weg in eine andere Moderne*. Frankfurt: Suhrkamp.

Beekman, Volkert. 2004. Sustainable development and future generations. *Journal of Agricultural and Environmental Ethics* 17 (1): 3–22.

Bongiovanni, Rodolfo, and Jess Lowenberg-DeBoer. 2004. Precision agriculture and sustainability. *Precision Agriculture* 5 (4): 359–387.

Borlaug, Norman E. 2007. Sixty-two years of fighting hunger: Personal recollections. *Euphytica* 157 (3): 287–297.

Bragdon, Susan H. 2016. Reinvigorating the public sector: The case of food security, small-scale farmers, trade and intellectual property rules. *Development* 59 (3–4): 280–291.

Britz, Johannes J., and Tomas A. Lipinski. 2001. Indigenous knowledge: A moral reflection on current legal concepts of intellectual property. *Libri* 51 (4): 234–246. https://doi.org/10.1515/libr.2001.234.

Buchanan, Allen, Tony Cole, and Robert O. Keohane. 2011. Justice in the diffusion of innovation. *Journal of Political Philosophy* 19 (3): 306–332. https://doi.org/10.1111/j.1467-9760.2009.00348.x.

Canavan, Chelsey R., Ramadhani A. Noor, Christopher D. Golden, Calestous Juma, and Wafaie Fawzi. 2017. Sustainable food systems for optimal planetary health. *Transactions of the Royal Society of Tropical Medicine and Hygiene* 111 (6): 238–240.

Carroll, La Shun L. 2017. A comprehensive definition of technology from an ethological perspective. *Social Sciences* 6 (4): 126.

Clague, Julie. 2006. "Patent Injustice": Applying Sen's capability approach to biotechnologies. In *Transforming unjust structures the capability approach*, ed. Séverine Deneulin, Mathias Nebel, and Nicholas Sagovsky, 177–196. Dordrecht: Springer.

Cline, William R. 2007. *Global warming and agriculture: Impact estimates by country*. Washington, DC: Center for Global Development.

COMEST. 2015. *Ethical Perspective on science, technology and society: A contribution to the post-2015 Agenda*. Paris: UNESCO & COMEST.

Coolsaet, Brendan. 2016. Towards an agroecology of knowledges: Recognition, cognitive justice and farmers' autonomy in France. *Journal of Rural Studies* 47: 165–171.

De Jonge, Bram. 2009. Plants, genes and justice: An inquiry into fair and equitable benefit-sharing. Ph.D. diss., Applied Philosophy Group, Wageningen University.

De Schutter, Olivier. 2009a. *International trade in agriculture and the right to food*. Geneva: Friedrich-Ebert-Stiftung.

De Schutter, Olivier. 2009b. Seed policies and the right to food: enhancing agrobiodiversity and encouraging innovation (Report presented to the UN General Assembly, 64th session, UN doc. A/64/170). New York: United Nations.

De Schutter, Olivier. 2017. The political economy of food systems reform. *European Review of Agricultural Economics* 44 (4): 705–731.

De Winter, Jan. 2012. How to make the research agenda in the health sciences less distorted. *THEORIA. Revista de Teoría, Historia y Fundamentos de la Ciencia* 27 (1): 75–93.

DeCamp, Matthew Wayne. 2007. Global health: A normative analysis of intellectual property rights and global distributive justice. Ph.D. diss., Department of Philosophy, Duke University.

Doran, Elizabeth M.B., Jay S. Golden, and B.L. Turner. 2017. From basic research to applied solutions: are two approaches to sustainability science emerging? *Current Opinion in Environmental Sustainability* 29: 138–144.

Drahos, Peter, and John Braithwaite. 2001. Intellectual property, corporate strategy, globalisation: TRIPS in context. *Wisconsin International Law Journal* 20: 451–480.

Drahos, Peter, and John Braithwaite. 2003. *Information feudalism: Who owns the knowledge economy?*. New York: New Press.

Drugs for Neglected Diseases Working Group. 2001. *Fatal Imbalance: The Crisis in Research and Development for Drugs for Neglected Diseases*. Geneva: Médecins Sans Frontières.

Dumitru, Speranta. 2008. Are Rawlsians entitled to monopoly rights? In *Intellectual Property and Theories of Justice*, edited by Axel Gosseries, Alain Strowel and Alain Marciano, 57–72. Houndmills & New York: Palgrave Macmillan.

Eastwood, Callum, Laurens Klerkx, Margaret Ayre, and Brian Dela Rue. 2019. "Managing socio-ethical challenges in the development of smart farming: from a fragmented to a comprehensive approach for responsible research and innovation." *Journal of Agricultural and Environmental Ethics*32: 741–768.

Eisenberg, Rebecca S., and Richard R. Nelson. 2002. Public vs. proprietary science: A fruitful tension? *Daedalus* 131 (2): 89–101.

Fernández Pinto, Manuela. 2015. Commercialization and the limits of well-ordered science. *Perspectives on Science* 23 (2): 173–191.

Figueroa, Robert, and Claudia Mills. 2001. Environmental justice. In *A companion to environmental philosophy*, edited by Dale Jamieson, 426–438. Malden & Oxford: Blackwell.

Franssen, Maarten, Gert-Jan Lokhorst, and Ibo van de Poel. 2018. Philosophy of Technology. In *The Stanford Encyclopedia of Philosophy (Fall 2018 Edition)*, edited by Edward N. Zalta.

Fraser, Nancy. 1998. Social justice in the age of identity politics: Redistribution, recognition, and participation. In *The tanner lectures of human values*, ed. Grethe B. Peterson, 1–67. Salt Lake City: The University of Utah Press.

Gault, Fred. 2012. User innovation and the market. *Science and Public Policy* 39: 118–128.

Gebbers, Robin, and Viacheslav I. Adamchuk. 2010. Precision agriculture and food security. *Science* 327 (5967): 828–831.

Glennie, Charlotte, and Alison Hope Alkon. 2018. Food justice: Cultivating the field. *Environmental Research Letters* 13 (7): 073003.

Gliessman, Stephen R. 2007. *Agroecology: The ecology of sustainable food systems*. Boca Raton: CRC Press.

Godin, Benoît. 2015. *Innovation contested: The idea of innovation over the centuries*. New York & Oxon: Routledge.

Goeschl, Timo, and Timothy Swanson. 2003. Pests, plagues, and patents. *Journal of the European Economic Association* 1 (2–3): 561–575.

Gomberg, Paul. 2007. *How to make opportunity equal*. New York: Wiley Blackwell.

Gosseries, Axel, Alain Marciano, and Alain Strowel (eds.). 2008. *Intellectual property and theories of justice*. Houndmills: Palgrave Macmillan.

Grunwald, Armin. 2009. Technology assessment: Concepts and methods. In *Philosophy of Technology and Engineering Sciences*, edited by Anthonie Meijers, 1103–1146. Amsterdam: Elsevier.

Halpert, Madeleine-Thérèse, and M Jahi Chappell. 2017. Prima facie reasons to question enclosed intellectual property regimes and favor open-source regimes for germplasm. *F1000Research* 6:284.

Hamilton, Neil D. 2008. Feeding the world's future: Agrarian justice and the rule of law. *Drake Journal of Agricultural Law* 13: 545–560.

Hassoun, Nicole. 2016. Individual responsibility for promoting global health: The case for a new kind of socially conscious consumption. *The Journal of Law, Medicine & Ethics* 44 (2): 319–331.

Hicks, Daniel J. 2017. Genetically modified crops, inclusion, and democracy. *Perspectives on Science* 25 (4): 488–520.

Hinsch, Wilfried. 2001. Global distributive justice. *Metaphilosophy* 32 (1–2): 58–78.

Hollis, Aidan, and Thomas W. Pogge. 2008. The health impact fund: Making new medicines accessible for all. In Oslo & New Haven: Incentives for Global Health. http://www.healthimpactfund.org/hif_book.pdf. Accessed 31 March 2012.

Holt-Giménez, Eric. 2002. Measuring farmers' agroecological resistance after Hurricane Mitch in Nicaragua: a case study in participatory, sustainable land management impact monitoring. *Agriculture, Ecosystems & Environment* 93 (1–3): 87–105.

Hoyningen-Huene, Paul. 2008. Systematicity: The nature of science. *Philosophia* 36 (2): 167–180.

ICSU Study Group on Science and Traditional Knowledge. 2002. Science and Traditional Knowledge: Report from the ICSU Study Group on Science and Traditional Knowledge.

IPBES. 2019. *Summary for policymakers of the global assessment report on biodiversity and ecosystem services of the Intergovernmental Science-Policy Platform on Biodiversity and Ecosystem Services*. Bonn: IPBES secretariat.

James, Aaron. 2012. *Fairness in practice: A social contract for a global economy*. New York: Oxford University Press.

Jefferson, Richard A. 2006. Science as social enterprise: The CAMBIA BiOS initiative. *Innovations: Technology, Governance, Globalization* 1 (4): 13–44.

Jenkins, Kirsten. 2018. Setting energy justice apart from the crowd: Lessons from environmental and climate justice. *Energy Research & Social Science* 39: 117–121.

Jenkins, Kirsten, Darren McCauley, Raphael Heffron, Hannes Stephan, and Robert Rehner. 2016. Energy justice: A conceptual review. *Energy Research & Social Science* 11: 174–182.

Jonas, Hans. 1979. *Das Prinzip Verantwortung: Versuch einer Ethik für die technologische Zivilisation*. Frankfurt am Main: Insel-Verlag.

Kaebnick, Gregory E. 2007. Putting concerns about nature in context: The case of agricultural biotechnology. *Perspectives in Biology and Medicine* 50 (4): 572–584.

Klerkx, Laurens, Barbara Van Mierlo, and Cees Leeuwis. 2012. Evolution of systems approaches to agricultural innovation: Concepts, analysis and interventions. In *Farming Systems Research into the 21st century: The new dynamic*, ed. Ika Darnhofer, David Gibbon, and Benoît Dedieu, 457–483. Dordrecht: Springer.

Kloppenburg, Jack. 2005. *First the seed: The political economy of plant biotechnology*, 2nd ed. Madison: University of Wisconsin Press.

Koepsell, David. 2010. Back to basics: How technology and the open source movement can save science. *Social Epistemology* 24 (3): 181–190.

Koepsell, David. 2016. *Scientific integrity and research ethics: An approach from the ethos of science*. Cham: Springer.

Lacey, Hugh. 2012. Reflections on science and technoscience. *Scientiae studia* 10 (SPE): 103–128.

Lappé, Frances Moore, Jennifer Clapp, Molly Anderson, Robin Broad, Ellen Messer, Thomas Pogge, and Timothy Wise. 2013. How we count hunger matters. *Ethics & International Affairs* 27 (03): 251–259.

Lemley, Mark A. 2012. The myth of the sole inventor. *Michigan Law Review*, 709–760.

Leopold, Aldo. 1949. *A Sand County almanac: Sketches here and there*. New York: Oxford University Press.

Lever, Annabelle (ed.). 2012. *New frontiers in the philosophy of intellectual property*. Cambridge: Cambridge University Press.

Linares Salgado, Jorge. 2013. "Valoración social del riesgo tecnocientífico controversias sobre el desarrollo y la innovación." In *Aproximaciones a la filosofía política de la ciencia*, edited by Carlos López-Beltrán and Ambrosio Velasco Gómez, 489–508. Mexico, DF: Universidad Nacional Autonoma de Mexico.

Lolas, Fernando (ed.). 2006. *Ética e innovación tecnológica*. Santiago: Centro Interdisciplinario de Estudios en Bioética.

Lolas, Fernando. 2014. Ética y conducta alimentaria: la obesidad y sus desafíos. *O Mundo da Saúde* 38 (3): 349–354.

Loo, Clement. 2014. Towards a more participative definition of food justice. *Journal of Agricultural and Environmental Ethics* 27 (5): 787–809.

Louwaars, Niels, Bram De Jonge, and Peter Munyi. 2013. Intellectual property rights in the plant sciences and development goals in agriculture: An historical perspective. In *Knowledge Management and Intellectual Property*, edited by Stathis Arapostathis and Graham Dutfield, 252–272. Cheltenham & Northampton: Edward Elgar.

Love, James, and Tim Hubbard. 2007. The big idea: Prizes to stimulate R&D for new medicines. *Chicago-Kent Law Review* 82 (3): 1519–1554.

Macnaghten, Phil, Richard Owen, Jack Stilgoe, R. Brian Wynne, A.De Azevedo, Jason Chilvers Campos, R. Dagnino, G. Di Giulio, and Emma Frow. 2014. Responsible innovation across borders: tensions, paradoxes and possibilities. *Journal of Responsible Innovation* 1 (2): 191–199.

Mazloumian, Amin, Dirk Helbing, Sergi Lozano, Robert P. Light, and Katy Börner. 2013. Global multi-level analysis of the 'Scientific Food Web. *Science Reports* 3 (1167). https://doi.org/10.1038/srep01167.

Mazoyer, Marcel, and Laurence Roudart. 2006. *A history of world agriculture: From the neolithic age to the current crisis.* New York: Monthy Review Press.

McIntyre, Beverly D., Hans R. Herren, Judi Wakhungu, and Robert T. Watson. 2009. *International assessment of agricultural knowledge, science and technology for development (IAASTD): synthesis report with executive summary: A synthesis of the global and sub-global IAASTD reports.* Washington, DC: Island Press.

McMichael, Anthony. 2017. *Climate change and the health of nations: Famines, fevers, and the fate of populations.* Oxford: Oxford University Press.

El Mujtar, Veronica, Nacira Muñoz, Barbara Prack Mc Cormick, Mirjam Pulleman, and Pablo Tittonell. 2019. Role and management of soil biodiversity for food security and nutrition; where do we stand? *Global Food Security* 20: 132–144.

Milanovic, Branko. 2011. *Worlds apart: Measuring international and global inequality*: Princeton University Press.

Nicholls, Clara I, and Miguel A Altieri. 2018. *Caminos para la amplificación de la agroecología.* Medellín: Centro Latinoamericano de Investigaciones Agroecológicas.

Nyéléni Forum for Food Sovereignty. 2007. *Declaration of Nyéléni.* Sélingue: Nyéléni Forum for Food Sovereignty.

Ooms, Gorik. 2010. Why the West is perceived as being unworthy of cooperation. *Journal of Law, Medicine and Ethics* 38 (3): 594–613. https://doi.org/10.1111/j.1748-720X.2010.00514.x.

Outterson, Kevin. 2014. New business models for sustainable antibiotics. In *Centre on Global health Security Working Group Papers, Working Groups on Antimicrobial Resistance, Paper 1.* London: Chatham House (The Royal Institute of International Affairs).

Owen, Richard, Phil Macnaghten, and Jack Stilgoe. 2012. Responsible research and innovation: From science in society to science for society, with society. *Science and public policy* 39 (6): 751–760.

Paine, Thomas. 1797. "Agrarian justice." Baltimore: Social Security Administration. https://www.ssa.gov/history/paine4.html.

Pant, Laxmi Prasad. 2019. Responsible innovation through conscious contestation at the interface of agricultural science, policy, and civil society. *Agriculture and Human Values* 36: 183–197.

Papaioannou, Theo. 2018. *Inclusive innovation for development: meeting the demands of justice through public action.* London: Routledge.

Patel, Raj, and Jason W. Moore. 2017. *A history of the world in seven cheap things: A guide to capitalism, nature, and the future of the planet.* Oakland, CA: University of California Press.

Patterson, James J., Thomas Thaler, Matthew Hoffmann, Sara Hughes, Angela Oels, Eric Chu, Aysem Mert, Dave Huitema, Sarah Burch, and Andy Jordan. 2018. Political feasibility of 1.5 C societal transformations: The role of social justice. *Current Opinion in Environmental Sustainability* 31: 1–9.

Pesch, Udo, Aad Correljé, Eefje Cuppen, and Behnam Taebi. 2017. Energy justice and controversies: Formal and informal assessment in energy projects. *Energy Policy* 109: 825–834.
Piketty, Thomas. 2014. *El capital en el siglo XXI*. Mexico, DF: Fondo de Cultura Económica.
Pogge, Thomas W. 2008. *World poverty and human rights: Cosmopolitan responsibilities and reforms*. 2nd ed. Cambridge: Polity.
Popp, József, Károly Pető, and János Nagy. 2013. Pesticide productivity and food security. A review. *Agronomy for Sustainable Development* 33 (1): 243–255.
Pretty, Jules. 2008. Agricultural sustainability: Concepts, principles and evidence. *Philosophical Transactions of the Royal Society of London B: Biological Sciences* 363 (1491): 447–465.
Ravn, Tine, Mathias Wullum Nielsen, and Niels Mejlgaard. 2015. Metrics and Indicators of Responsible Research and Innovation: Progress report D3.2. Brussels: European Commission.
Reiss, Julian. 2010. In favour of a Millian proposal to reform biomedical research. *Synthese* 177: 427–447.
Resnik, David B. 2009. International standards for research integrity: An idea whose time has come? *Accountability in research* 16 (4): 218–228.
Richards, Paul. 1993. Cultivation: Knowledge or performance? In *An anthropological critique of development: The growth of ignorance*, ed. Mark Hobart, 73–90. London: Routledge.
Rip, Arie. 2014. The past and future of RRI. *Life sciences, society and policy* 10: 17.
Robaey, Zoë, and Cristian Timmermann. 2018. Fair agricultural innovation for a changing climate. In *Food Justice, the Environment, and Climate Change*, edited by Erinn Cunniff Gilson and Sarah Kenehan, 213–230. Lanham: Rowman & Littlefield.
Rodney, Walter. 1972. *How Europe underdeveloped Africa*. London: Bogle-L'Ouverture Publications.
Röling, Niels. 2009. Pathways for impact: Scientists' different perspectives on agricultural innovation. *International journal of agricultural sustainability* 7 (2): 83–94.
Rosset, Peter M., and Miguel A. Altieri. 1997. Agroecology versus input substitution: A fundamental contradiction of sustainable agriculture. *Society & Natural Resources* 10 (3): 283–295.
Ruger, Jennifer Prah. 2015. Governing for the common good. *Health Care Analysis* 23 (4): 341–351.
Schlosberg, David, and David Carruthers. 2010. Indigenous struggles, environmental justice, and community capabilities. *Global Environmental Politics* 10 (4): 12–35.
von Schomberg, René. 2012. Prospects for technology assessment in a framework of responsible research and innovation. In *Technikfolgen abschätzen lehren*, ed. Marc Dusseldorp and Richard Beecroft, 39–61. Wiesbaden: VS Verlag für Sozialwissenschaften.
von Schomberg, René. 2015. Responsible innovation: The new paradigm for science, technology and innovation policy. In *Responsible Innovation: Neue Impulse für die Technikfolgenabschätzung*, ed. A. Bogner, M. Decker, and M. Sotoudeh, 47–70. Nomos: Baden-Baden.
Schroeder, Doris, and Balakrishna Pisupati. 2010. *Ethics, justice and the convention on biological diversity*. Nairobi: United Nations Environmental Program.
Sen, Amartya. 1981. *Poverty and famines. An essay on entitlement and deprivation*. Oxford & New York: Oxford University Press.
Shrader-Frechette, Kristin. 2002. *Environmental justice: Creating equality, reclaiming democracy*. Oxford: Oxford University Press.
Shue, Henry. 1996. *Basic rights: subsistance, affluence, and U.S. foreign policy*, 2nd ed. Princeton, N.J.: Princeton University Press.
Siipi, Helena. 2008. Dimensions of naturalness. *Ethics and the Environment* 13 (1): 71–103.
Springmann, Marco, Michael Clark, Daniel Mason-D'Croz, Keith Wiebe, Benjamin Leon Bodirsky, Luis Lassaletta, Wim de Vries, Sonja J Vermeulen, Mario Herrero, and Kimberly M Carlson. 2018. Options for keeping the food system within environmental limits. *Nature* 562: 519–525.
Stilgoe, Jack, Richard Owen, and Phil Macnaghten. 2013. Developing a framework for responsible innovation. *Research Policy* 42 (9): 1568–1580.
Sunder, Madhavi. 2012. *From Goods to a Good Life: Intellectual Property and Global Justice*: Yale University Press.

Thompson, Paul B. 2017. *The spirit of the soil: Agriculture and environmental ethics*. New York: Routledge.

Tierney, William G., and Karri A. Holley. 2008. Inside Pasteur's quadrant: knowledge production in a profession. *Educational Studies* 34 (4): 289–297.

Timmermann, Cristian. 2013. *Life sciences, intellectual property regimes and global justice*. Ph.D. diss., Philosophy Group, Wageningen University.

Timmermann, Cristian. 2014a. An assessment of prominent proposals to amend intellectual property regimes using a human rights framework. *La Propiedad Inmaterial* 18: 221–253.

Timmermann, Cristian. 2014b. Sharing in or benefiting from scientific advancement? *Science and Engineering Ethics* 20 (1): 111–133.

Timmermann, Cristian. 2018. Contributive justice. An exploration of a wider provision of meaningful work. *Social Justice Research* 31 (1): 85–111. https://doi.org/10.1007/s11211-017-0293-2.

Timmermann, Cristian, Georges F. Félix, and Pablo Tittonell. 2018. Food sovereignty and consumer sovereignty: Two antagonistic goals? *Agroecology and Sustainable Food Systems* 42 (3): 274–298. https://doi.org/10.1080/21683565.2017.1359807.

Timmermann, Cristian, and Zoë Robaey. 2016. Agrobiodiversity under different property regimes. *Journal of Agricultural and Environmental Ethics* 29 (2): 285–303. https://doi.org/10.1007/s10806-016-9602-2.

Tittonell, Pablo. 2013. *Farming systems ecology: Towards ecological intensification of world agriculture*. Wageningen: Wageningen Universiteit.

Tittonell, Pablo. 2016. Feeding the world with soil science: embracing sustainability, complexity and uncertainty. *SOIL Discuss.* https://doi.org/10.5194/soil-2016-7.

Tittonell, Pablo, Laurens Klerkx, Frederic Baudron, Georges F. Félix, Andrea Ruggia, Dirk van Apeldoorn, Santiago Dogliotti, Paul Mapfumo, and Walter A.H. Rossing. 2016. Ecological intensification: local innovation to address global challenges. *Sustainable Agriculture Reviews* 19: 1–34.

Torrance, Andrew W., and Eric von Hippel. 2015. The right to innovate. *Michigan State Law Review* 2015 (2): 793–829.

UN Committee on Economic Social and Cultural Rights. 1999. *General Comment No. 12. The right to adequate food (article 11) (E/C.12/1999/5)*. Geneva: United Nations Economic and Social Council.

UN Committee on Economic Social and Cultural Rights. 2006. General Comment No. 17: The Right of Everyone to Benefit from the Protection of the Moral and Material Interests Resulting from any Scientific, Literary or Artistic Production of Which He or She is the Author (Art. 15, Para. 1 (c) of the Covenant, E/C.12/GC/17). Geneva: United Nations Economic and Social Council.

van den Belt, Henk. 2010. Robert Merton, Intellectual Property, and Open Science. In *The commodification of academic research: Science and the modern university*, ed. Hans Radder, 187–230. Pittsburgh: University of Pittsburgh Press.

van Est, Rinie, Bart Walhout, and Frans Brom. 2012. Risk and technology assessment. In *Handbook of risk theory: Epistemology, decision theory, ethics, and social implications of risk*, ed. Sabine Roeser, Rafaela Hillerbrand, Per Sandin, and Martin Peterson, 1067–1091. Dordrecht: Springer.

Van Haperen, P.F., B. Gremmen, and J. Jacobs. 2012. Reconstruction of the ethical debate on naturalness in discussions about plant-biotechnology. *Journal of Agricultural and Environmental Ethics* 25 (6): 797–812.

Vanloqueren, Gaëtan, and Philippe V. Baret. 2009. How agricultural research systems shape a technological regime that develops genetic engineering but locks out agroecological innovations. *Research Policy* 38 (6): 971–983.

Via Campesina. 1996. *The right to produce and access land*. Rome: Via Campesina.

Wezel, Alexander, Stéphane Bellon, Thierry Doré, Charles Francis, Dominique Vallod, and Christophe David. 2009. Agroecology as a science, a movement and a practice. A review. *Agronomy for Sustainable Development* 29 (4): 503–515.

Williams, Justine M., and Eric Holt-Giménez. 2017. *Land justice: Re-imagining land, food, and the commons*. Oakland: Food First Books.
Wise, Timothy A. 2019. *Eating tomorrow: agribusiness, family farmers, and the battle for the future of food*. New York: The New Press.
Wynberg, Rachel, Doris Schroeder, and Roger Chennells (eds.). 2009. *Indigenous peoples, consent and benefit sharing: Lessons from the San-Hoodia case*. Dordrecht: Springer.
Ziegler, Jean. 2011. *Destruction massive: Géopolitique de la faim*. Paris: Seuil.
Zwart, Hub, Laurens Landeweerd, and Arjan van Rooij. 2014. Adapt or perish? Assessing the recent shift in the European research funding arena from 'ELSA'to 'RRI'. *Life Sciences, Society and Policy* 10: 11.

Chapter 2
Intellectual Property Regimes and Their Impact on Agricultural Research and Development

Abstract This chapter gives an overview of the different intellectual property regimes that cover innovations in agriculture. It starts by giving an overview of the main justifications for intellectual property. It continues by discussing philosophical arguments based on desert, need and personality theories. After that, it offers an overview of the main forms of intellectual property protection, concentrating on patents, plant breeders' rights, copyright and geographical indications. Finally, it introduces some of the major problems of proprietary science, in relation to the six dimensions of social justice.

Keywords Patents · Plant breeders' rights · Incentive arguments · Desert · Moral interests

2.1 Introduction

As is widely known, intellectual property has acquired a central role in incentivizing research and development. Agricultural innovation, independently of whether it is pursued in public institutions, universities, or industry, is no exception. In this chapter, I discuss the effect of the use and existence of intellectual property rights on agricultural innovation, and how this affects social justice. To do so, I concentrate on two types of intellectual property regimes: patents and plant breeders' rights, and provide some additional remarks on the effects of copyrights and protected geographic indications. As this topic has been widely and sufficiently dealt with in the existing literature (Prifti 2015; Haugen et al. 2011; Correa 1995), I just offer a brief account here, so as to allow us to better understand the significance of intellectual property regimes for agricultural innovation, and the opportunities and conflicts the wide use and the very existence of such exclusive rights create for social justice.

In this chapter, I start by explaining what intellectual property rights are and how they are justified. Then I introduce the main intellectual property regimes that affect agricultural innovation. After that I proceed to the question: what are the benefits and costs for society and the environment of overly relying on proprietary agricultural science to come up with technological solutions?

C. Timmermann, *Social Justice and Agricultural Innovation*, The International Library of Environmental, Agricultural and Food Ethics 31,
https://doi.org/10.1007/978-3-030-56193-2_2

2.2 What Are Intellectual Property Rights and What Is Their Purpose?

In general terms, intellectual property rights are temporary exclusive rights over inventions, designs or expressions. The basic idea behind these rights is to offer creative minds temporary exclusivity over the objects of their innovations, so that they may recoup the financial costs of research and development, earn profits and gain recognition for their efforts, on the condition that they disclose their innovations and the knowledge needed to reproduce them. Once the period of exclusivity elapses, the knowledge encompassing the invention or the form of expression enters the public domain, free to be used unless otherwise stipulated by law. This social contract, offering temporary exclusivity in return for disclosing the methods for reproducing the invention, is often called "the patent bargain" (Biagioli 2006; Liivak and Peñalver 2013).

Innovators are granted the right to exclude others from using their inventions for a standardized period of time; in the case of patents this is usually 20 years. The time of protection is completely independent of any social need for the invention, or the effort dedicated to its development. The granting of a patent does not imply that innovators are automatically entitled to use the invention themselves: national regulations, in particular biosafety laws, may limit or withdraw usage rights.

Since intellectual property are exclusive rights, by their nature they will have the effect that some people, for a variety of reasons or circumstances, but primarily due to poverty, end up without access. As this can have a substantial effect on social welfare, we proceed by analysing the different justifications of intellectual property in terms of their effects on social justice. We can name three major philosophical traditions that defend the existence of intellectual property regimes: (i) arguments claiming that innovators deserve these rights as a fair reward for their intellectual labour, (ii) utilitarian justifications arguing that these rights are necessary to speed up and allow investments in innovation and thereby improve social welfare, and (iii) the claim that these rights allow the protection of special interests of the innovator, based on so-called "personality theories" of property. Let us review each of these justifications.

2.2.1 Intellectual Property: Entitlement and Deservingness

A very common defence of intellectual property is to argue that innovators deserve a fair reward for their intellectual labour. This has led to a situation where ownership of the fruits of one's labour is directly appealed to, by reference to notions of desert. An immediate problem with desert arguments is that there are different understandings of "desert" (De Jonge 2011). We can understand desert in a very loose sense, as something that would be nice to provide if circumstances allow it. In a stricter sense, however, one could claim that creative minds not only deserve a fair reward, but are actually entitled to the fruits of their labour. As we will see, this additional step has substantial effects when arguing for social interests in innovations, particularly the human right to benefit from scientific advancement. Let us review the common intellectual history of these demands.

The idea of understanding intellectual property rights as a fair reward draws strongly on traditional ideas that were originally conceived to justify property rights over land, in particular John Locke's theory of property (Drahos 1996). The philosopher Locke famously claimed that one can legitimately assert property rights over something by mixing one's labour with it. Yet from the very beginning, this assertion was subject to explicit conditions or provisos: the object must be unowned, that is, nobody has previously claimed it; there is "enough and as good" left for others; and the enclosure (or appropriation) should not lead to wastage (Locke 1689). While there is some ambiguity concerning their interpretations and scope, the latter two conditions are usually referred to in the literature as the two "Lockean provisos".

There are different justifications behind these provisos. To be able to benefit from work—something that is intrinsically ours, since we own our bodies (an argument backed by philosophical principles of self-ownership)—it is crucial to have property rights over our work, otherwise others will harvest the fruits of our efforts. Here it is crucial that labour implies an intention to achieve a particular outcome, and thereby requires knowledgeable action (Cwik 2014). In this tradition we can also find a utilitarian argument: society also wants to create a certain incentive for people to mix their labour with things that can be improved, since this will be to the benefit of all (Strauss 1952).

A less recognized but well-documented interpretation claims that Locke had a third proviso in mind: supporting charity, in the sense that everyone should have access to goods to secure their subsistence (Widerquist 2010). Recent scholarship shows that the idea of necessity—sometimes even defended as a right of necessity—has the moral strength to overturn other rights, especially property rights, and was supported by some prominent thinkers, both during Locke's lifetime and earlier (Mancilla 2012; O'Neill 2018).

According to this line of thought, as long as these provisos are respected, by mixing their labour with it, people deserve ownership of the improved object. Translating this argument to the realm of intellectual goods, however, brings in additional ethical issues. While there is a necessary link between being able to enjoy the fruits of one's labour and having continuous access to the tangible object with which one has mixed one's labour, being able to use an object of innovation does usually not require exclusivity (Timmermann 2017). Knowledge is a public good and thus of non-rivalrous consumption, meaning that it can be enjoyed by multiple people at the same time without exhausting it (Stiglitz 1999). Or, as Thomas Jefferson (1813) famously stated in a more eloquent form: "He who receives an idea from me, receives instruction himself without lessening mine; as he who lights his taper at mine, receives light without darkening me." Thus what is really at stake is not the freedom to enjoy the fruits of one's labour, but rather the possibility of earning higher revenues by exclusively selling one's invention, when others are willing and able to pay for it (Wilson 2009; Gosseries 2011). Here it is important to be aware that other forms of securing the efforts of one's intellectual labour are conceivable, such as prizes and direct government funding (Hettinger 1989; Love and Hubbard 2007). The absence of an alternative funding mechanism is a moral failing in itself, and not a morally acceptable justification for excluding others from the benefits of scientific advancement, particularly when such exclusion jeopardizes the securing of basic needs (Timmermann 2014a). In other words, an inability to benefit exclusively from the fruits of one's labour does not morally justify hindering others from fulfilling people's basic needs.

Moreover, the idea that innovators are tapping into an endless pool of resources is not only questioned in relation to tangible objects, but also in relation to intellectual resources. Innovators do not usually get their ideas out of the blue, but rather from a body of emerging knowledge. Ideas for innovation taken exclusively from this body are methods and tools that others will consequently be unable to develop independently. In view of our current state of knowledge there are only a limited number of ways to develop a distinctive technological solution (Trerise 2010). This situation is particularly problematic for the patents involving genetic resources in which innovators cannot invent around (Van Overwalle 2010).

In addition, a number of objections to the idea of justifying exclusivity for inventions by appealing to notions of desert can be found in the literature.

First, the economic rewards that one may gain from a patent do not adequately reflect the effort one has put into it (Sterckx 2006) or the social utility of the invention. Intellectual property only rewards certain types of labour, leaving other types of creative work outside its incentive system (Sherman 2015). The economic reward depends on market demand and the possibility of selling one's invention. Especially when consumers are in dire need of the invention, patent holders can engage in highly exploitative rent-seeking practices that allow an extraction of capital that bears no relation to the amount of labour they devoted to the invention—or the money they spent acquiring the patent (Timmermann 2014b). Conversely, this incentive system does not stimulate the development of easily replicable inventions, even when they have a high social utility, as innovators may not recoup their research and development costs through sales. Also, it insufficiently stimulates research on products meant to solve the problems of poor consumers, even when this would enormously improve their welfare (Hassoun 2015).

Second, especially in the case of patents and plant breeders' rights, those applying for intellectual property protection rely on a vast amount of previous work done by others (Hettinger 1989). Moreover, intellectual property assumes that inventors can separate their own labour from other contributions that were decisive for bringing the invention into existence, something that is highly questionable (Radder 2013). The system also fails to distinguish between deserving *a reward* and deserving *the full reward* from creative work (Papaioannou 2006; Attas 2008). Furthermore, as a winner-takes-all system, intellectual property overly benefits those who are lucky enough to be able to arrive first at the patent office with an intellectual contribution that fits the relevant legal requirements. This is particularly problematic and unjust for independent inventors. For example, those who independently come up with the same invention the day after the patent was filled are not even allowed to use the invention themselves without a license, ultimately depriving them of their fruits of their own labour (Nozick 1974; Varelius 2014). There is no easy solution to this problem. In practice, when the stakes are high, it becomes extremely difficult to prove independent invention, since research plans can only show that one came close to an invention, and not that one also solved the final problems involved in developing a ready product or process (Breakey 2009).

Third, the idea that innovators deserve intellectual property, and furthermore that governments should protect such rights, implies that innovation generally has a positive nature. In principle, international intellectual property treaties foresee that governments may want to restrain from awarding exclusive rights to inventions that are morally objectionable, in the sense of being contrary to public morals (Tvedt and Forsberg 2017). In practice, however, governments exercise this liberty very narrowly, mostly in relation to human cloning techniques (Conde Gutiérrez 2017). In contrast, governments have allowed exclusive rights to agrochemical compounds that can be developed into substances that are primarily used as weapons for attacking food production facilities, such as agent orange (Jacob and Walters 2005). Even when we do not have such extreme cases in mind, it becomes clear that when we talk about giving something valuable as a desert, we want to reward socially beneficial conduct. Yet patent offices rarely make objections on such grounds (Prifti 2016).

The very framing of these exclusive rights as intellectual property has also been criticized (Heins 2008). By understanding material interests as property rights instead of privileges, these interests gain the additional protection that is commonly assigned to material property. This change of language is not only linguistically significant, but also comes with some major intellectual baggage (Chopra 2018), as material property rights were traditionally recognized in order to protect vital assets such as shelter, means of production and privacy. The concept of property is closely linked to the idea of being entitled to (and thus deserving) full exploitation: something that the past idea of privilege does not entail.

A benefit of this system is that it allows markets to set the direction of research. Governments may have insufficient information either to set research agendas without bias and favouritism, or to identify adequate reward sizes (Belleflamme 2006). Moreover, research expenses will in large part be covered by the consumers who decide to pay for the developed products. In contrast with taxation, where everyone contributes towards the development of technologies, with intellectual property the people who make use of innovations are the only ones who have to pay innovators. People favouring market solutions often claim that people who cannot afford the vital innovations they need could be assisted with vouchers or government programmes. Moreover, by building a strong link between desert and invention, inventors develop a strong attachment to their inventions, and—if they are altruistic or even just ambitious—also a strong interest in *their* invention having a positive impact, thus incentivizing them to make sure that the invention is widely accessible.

2.2.2 Needing Intellectual Property

The second most common theoretical approach to justifying intellectual property is rule utilitarianism; in the sense that following the general rule of awarding exclusive rights to innovators yields the highest social welfare improvement. This line of reasoning can already be seen in the United States Constitution (1789), where the Congress is empowered to "promote the Progress of Science and useful Arts, by securing for limited Times to Authors and Inventors the exclusive right to their respective Writings and Discoveries" (art. I, §8, clause 8). Here intellectual property is meant to serve the specific policy goal of promoting science and the arts (Waldron 1992). The underlying argument is that creative minds and their investors will be encouraged to invest more resources in research and development if they have the prospect of securing exclusive rights for use of the object of innovation. Moreover, there should be clear guidelines on which types of invention are eligible for these rights, as this will increase certainty when making long-term investments in research and development.

A second argument for intellectual property, also referring to its instrumental nature, is to claim that the legal obligation to include all the information needed to replicate the invention benefits society at large, by expanding the amount of knowledge that will be available in the public domain once the patent expires (Sterckx 2006). In a world without intellectual property, trade secrecy will be seen as the key to exclusivity, and this may slow down scientific advancement as researchers will have a strong incentive to hide information that could facilitate the replication of an invention. Intellectual property would also provide a strong incentive to continue contributing to the development of inventions until they are ripe for the market (Sterckx 2011).

The reasoning behind these two claims relies on three major premises: (1) society should implement systems or institutions if these maximize social utility, (2) a system that confers limited and exclusive rights to authors and inventors is a necessary incentive to stimulate the production of intellectual works, and (3) encouraging innovations and their dissemination improves social welfare (Moore 2008). Again here, we can observe several problems with all of these assumptions.

To start, as we have seen, intellectual property is only one of many possible systems for encouraging innovation. The welfare-enhancing role of such a system needs to be compared to the alternatives to the existing system—for instance, increased government funding—and the absence of such systems. Yet as Fritz Machlup asserted many decades ago, one of the key challenges of intellectual property regimes is that we can hardly empirically verify the true advantages of such systems (Merges 2011; Trerise 2016; Lambrecht 2017), as we could hardly build a parallel world without intellectual property in order to draw comparisons. We are also owed evidence that shows that the best way to incentivize innovation is to have an intellectual property system whereby the same rules apply to all countries in the world, irrespective of their development status (James 2012).

There are also doubts concerning the second premise. The claim that people will stop innovating unless they can reap the full benefits of their creative efforts is not backed by either current or historical experience. Mathematics and agroecology, for instance, are fields that largely work without patents. The same goes for willingness to share one's inventions. As James Wilson (2010) notes, the fact that others may benefit from one's invention should not be perceived as a disincentive, for even when innovators are only moderately altruistic it actually counts as an incentive. We may recall a saying attributed to Benjamin Franklin—"Man is a tool-making animal"—to stress that it is part of human nature to innovate and build tools. At most, one could claim that we would have less innovation if innovators could not secure exclusive rights, but we would still have to prove that the advantages of this increased rate of innovation would outweigh the disadvantages of exclusivity. Another issue that needs to be considered, as will be seen in more detail in the next chapter, is that intellectual property also affects the types of innovation that will emerge and how science is done. To legitimize intellectual property, we would also need to prove that exclusivity leads to a better type of innovation.

From a moral perspective, the strongest argument against such systems is that they create artificial scarcity. The commercialization of innovations creates significant dead weights which translate, in the case of live-saving innovations, to literally dead weights, when people are too poor to pay for the goods they need (Pogge 2010). This means that, due to wide discrepancies in purchasing power, it is impossible to adjust sales practices in such a way that those who need access to the invention pay as much as they can (or are willing to), without leaving people empty-handed because of insufficient funds (Timmermann 2017). Intellectual property not only creates dead weights among end users, but also among the follow-up innovators—especially small- and medium-sized enterprises—who are impeded from using the invention to develop new products (Lewin 2007). Artificial scarcity is a major social justice issue. When crops fail because knowledge was withheld from those unable to pay the fees, this may lead to socially-avoidable cases of hunger and malnutrition (Wilson 2007). In terms of social cohesion, suffering from an avoidable harm is notably different from suffering from an unavoidable harm, making it much more difficult or even impossible to see the current social order as just. It is particularly difficult to find support among the general public for the cost of artificial scarcity caused by intellectual property, especially when this is linked to avoidable suffering (Forsberg et al. 2017; Hull 2009).

In the very early period, patent holders were required to make use of their patents in order to retain their exclusive rights. That is to say, inventors who kept their inventions inaccessible to the public lost their privileges (May 2007). Nowadays patent holders are not legally obliged to make their inventions accessible, unless specifically compelled to do so (Liivak and Peñalver 2013). This has led to a so-called "destructive" use of patents: a practice where patent holders use their exclusive rights to block further research or the marketing of products, especially when such strategies allow more lucrative competing products to be sold (Schneider 2010). Such practices can have strong negative effects on future research when patents are granted very broadly for research tools or foundational technologies (Feeney et al. 2018). It goes without saying that such behaviour goes against the spirit of patent laws and clearly undermines their utilitarian justification, as it hinders both the use of technological solutions and the possibility of improving the invention in question (Timmermann 2014b). For hard cases, governments reserve the right to impose compulsory licenses,[1] yet smaller countries in particular hardly ever make use of such rights.

A major advantage of this system is that inventors who come up with innovations with a strong potential market are much likelier to find investors who can help finance their research and development costs. The benefit of this is that inventors will not have to rely exclusively on public money, gifts or their own capital to finance their endeavours. Moreover, by establishing clearly-defined property rights and a clear record of who owns what, innovations can easily be licensed to whoever is willing to pay for them or offers something of interest in return. This allows, according to free market principles, a division of labour in which everyone can produce the good that they most effectively produce and exchange it with others. An effective innovator can thus license her invention to a more effective manufacturer. Lastly, it circumvents the "paradox of information" pointed out by Kenneth Arrow: if the inventor discloses her invention before she has a contract, then she has nothing left to enforce an offer, but if she does not disclose it, then the buyer has no idea what is being offered (Merges 1994). Patents can resolve this paradox by securing property rights.

Philosophically, relying on certain understandings of utilitarianism to justify intellectual property raises complications. If we argue that, by utilitarian calculations, patents lead to greater well-being, then we still need to justify why we should not deviate from this rule in cases where cancelling certain patents would lead to an even greater good (Resnik 2003). The very reasoning that justifies patents may be used to justify exceptions from patenting or compulsory licenses, which in practice would create substantial uncertainty about how much those investing in research and development could reasonably expect in financial returns (cf. Hsieh 2000).

2.2.3 Personality Theories and Intellectual Property

A third line of reasoning used to justify the existence of intellectual property is provided by personality theories. Such theories are inspired by the German philosopher Hegel, who claimed that we as individuals own our character traits, talents and feelings, and therefore have a strong interest in maintaining control over the objects that embody these traits (Moore 2011). Possessing property which incorporates traits of one's personality contributes to their extension, whenever others recognize such objects (Hughes 1988). According to this view, the creations of one's mind reflect part of the author's personality, so authors have a valid interest in ensuring that this image adequately reflects their personality (Moore 2008). Due to the strong link between the author of an invention and the object of invention, inventors should, by

[1]See Agreement on Trade-Related Aspects of Intellectual Property Rights (1995), art. 31.

this reasoning, have a right to decide how and if their invention is to be replicated: something that requires property rights.

In contemporary intellectual property laws, we find this reasoning reflected in two protected rights that are labelled as the "moral interests" of authors. These are the right to attribution of authorship and, in copyright, also the right to control the integrity of one's work. In ideal circumstances this means that authors have a right to be mentioned as such, and that they also have a right to object to modifications of their work, so as to ensure that their different depictions do not diverge from how the author originally intended them to be. To illustrate with an anecdote, as recounted in his memoirs, Gabriel García Márquez was so appalled by his publisher's decision to modify, in his first novel, the original Colombian Spanish of his characters to an Iberian Spanish—which distorted the Caribbean context and deprived it of its beauty—that he blocked the diffusion the novel, at the price of having to wait for years before seeing his work published again (García Márquez 2002). In practice however, lesser-known authors in particular have lost many of these traditional rights. It is common to grant publishers the right to make editorial modifications to one's work. The names of many important collaborators are seldom listed as inventors in patent applications. It is rare for hired scientists to retain any rights to object to any specific licensing practices concerning their work.

Moral rights are not uncontroversial. There are governmental measures that aim to protect creators, for example, limiting the alienability of the right to attribution of authorship, and especially for visual artists in some jurisdictions, granting even a right to a percentage of the income from subsequent sales of their work (Beitz 2005). In some cases, such measures are grounded in paternalism, due to the vulnerable position of many creators or inventors at an early-career stage, while in other cases, there is a strong social interest in limiting ghost-writing.

An important difference between the moral and economic interests of authors is that the former do not expire. Authors retain the right to be attributed as authors even after the protection of their economic interests expires. Good editorial practice also recognizes that edited versions should be labelled as such.

Personality-based theories offer only a very weak argument for limiting the diffusion of innovation in agriculture (Timmermann 2014b). They do, however, reveal the strong tradition of justifying the right to attribution of authorship. As will be discussed later (in Chaps. 6 and 9), failing to recognize people or communities as innovators violates fundamental interests.

2.3 Varieties of Intellectual Property Rights

That people should have the opportunity to secure the material and moral interests of their inventions is a right recognized in human rights charters and constitutions. These interests have been materialized all over the world by the recognition of intellectual property. Since 1996, the Agreement on Trade-related Aspects of Intellectual Property Rights (TRIPS) has enforced minimum protection guarantees for such rights, by linking intellectual property to trade agreements (Kapczynski 2008; James 2014). Thus non-complying countries can be commercially sanctioned (Drahos and Braithwaite 2003). Intellectual property became much more central since the 1980s, because public institutions and non-profit organizations, such as universities and governmental research institutes, became very interested in seeking exclusive rights in order to gain additional revenues (Radder 2010; Stephan 2012). Much science is now only pursued if it leads to products that can be retained exclusively, and this has distorted the traditional divide between public and private research. Nowadays, the public sector spends a few percentage points less than the private sector on research in higher-income countries, while it still makes the highest investments in middle-income countries (Anderson 2018). However, as public-sector and university scientists are pushed to seek industry collaboration, the private sector is gaining more and more say on the direction of science and how research should be done, despite the fact that much of their research money continues to come from taxpayers. Such influence is incompatible with democratic decision-making over research agendas, and poses many problems that will be discussed in later chapters.

The possibility of applying for intellectual property rights, as well as the type of exclusivity one may seek, depends on the type of agricultural innovation one develops and the global state of the art. Let us briefly review the main forms of intellectual property protection that affect agricultural innovation.

2.3.1 Patents

Patents have received most criticism in relation to intellectual property. According to the standards set by the Agreement on Trade-Related Aspects of Intellectual Property Rights (TRIPS), in order to apply for a patent, inventions must meet three key criteria: be new, involve an inventive step, and be capable of industrial application.[2] For an invention to count as novel, it is crucial that it has not been made public before, irrespective of any formal or informal recordings. If patent examiners find the invention to be documented in any form with a date prior to the patent application, they can lawfully invalidate the application. The invention must also be novel, in the sense that the inventor conceived something that did not previously exist in nature (Resnik 2016). A second requirement is that the invention involves an inventive step, or as some jurisdictions specify, is non-obvious to someone skilled in the art. Here global standards are taken as the norm, which means that the same standards of novelty apply for inventors of low-income countries, as for those of high-income countries. The bar for what is considered obvious is also set higher over time: what is considered inventive now may be regarded as obvious once it becomes common knowledge among experts. And third, the invention should have industrial applicability, which usually requires the invention to be replicable in larger scales. This demands that inventions be specified in precise ratios and that they do not require too

[2]See Agreement on Trade-Related Aspects of Intellectual Property Rights (1995), art. 27.1.

much tacit knowledge to replicate. Those who are granted a patent gain the right to exclude others from using their inventions for a period of time: 20 years in most jurisdictions.[3] However, as mentioned before, a patent does not automatically entitle its owner to use her invention, and this is of special concern when biosafety regulations may hinder use.

Patents have to be sought in the different countries in which one seeks protection through formal applications. Fees are generally requested for the application and to maintain the validity of patents (Landes and Posner 2004). Inventors must provide a patent document containing all the information needed to duplicate their invention and translate this document into the official languages of each country in which they request patent protection: something for which they usually have to hire specially-qualified attorneys and translators. Due to these requirements, a patent that covers major markets may involve major costs, shifting large amounts of scarce resources away from science to the legal apparatus needed to secure exclusive rights (Korthals and van den Belt 2014). In addition, the patent system has made it extremely profitable for companies that do not engage in research to manage other people's patents and litigate them in courts, often by exploiting the ambiguity of the language used in patent documents and turning uncertainty about the boundaries of patent claims to their advantage. This comes at a massive financial cost to innovators, and little of the money earned through these means flows back to innovation (Bessen et al. 2011). The patent system drains, through multiple channels, large amounts of financial resources and professional expertise from research, as it involves massive administrative costs.

Patents can also be used to avoid critical studies of technology. For example, in the case of GM seeds, researchers need to use this technology in order to be able to test its biosafety, and to perform control trials under different conditions in order to check if the claims of the patent holders meet commercial standards (Biddle 2014). Yet patent holders have substantial liberty to oppose or at least postpone such trials, by making full use of their exclusive rights. Every success in postponing such trials is of considerable advantage to patent holders, as they can continue to make full use of their exclusive rights, and thereby continue to make profits.

In some cases, it would be pointless to patent. Easily replicable inventions are rarely patented, since anyone could copy the invention in their own farm without paying royalties, and it would be prohibitively expensive to monitor this.

In agriculture, it would be a mistake to associate patenting solely with biotechnology. Besides biomarkers and genetically modified organisms, inventions in urban farming, food preservation, tools, machinery, pesticides and fertilizers—among other things—may also qualify for patents. For example, patentable innovations in urban farming, contributing to the development of automation and control sensors, have great potential for improving the efficient use of resources and reducing food miles.

In general terms, many of the negative effects of patents can be reduced by socially responsible licensing. Some companies and universities are already implementing and exploring humanitarian licensing strategies that waive fees for users with low purchasing power, or have differential pricing mechanisms that widen access to

[3] See Agreement on Trade-Related Aspects of Intellectual Property Rights (1995), art. 33.

consumers with fewer financial resources. The initiative to make Golden Rice—a rice enriched with ß-carotene—widely available, is the most prominent example of pooling together patents for humanitarian use (Van Overwalle 2009).

2.3.2 *Plant Breeders' Rights*

A distinctive feature of plant breeding is that it has directly depended for centuries on distant and diverse input for improving crop varieties. The earliest efforts to improve crops consisted in reseeding varieties with the most desired traits for the next harvest. As people started to move over vast territories, some brought in new varieties and species from different areas of the world. The varieties that were brought in from different regions were crossed, sometimes successfully and sometimes not, with local varieties. Some newly introduced plant species managed to survive and even prosper in their new environments. Until the late nineteenth century—that is, before Mendelian Laws were rediscovered—farmers and plant breeders did not understand the basics of heredity, and instead relied on random plant crossing in order to later select those varieties with the most attractive traits (Kloppenburg 2005). Nevertheless, from early on, farmers were aware of the importance of acquiring new varieties and testing them in their fields, dedicating substantial time to this effort. Also, governments and private institutions invested a large amount of resources in building botanical gardens and seed banks in order to collect and study beneficial and lucrative varieties (Schiebinger 2004; Moore 2018). Nowadays both private and public institutions keep enormous seed banks with varieties from all over the world to facilitate plant breeding, and as repositories to conserve agrobiodiversity against losses due to massive pests or catastrophic events (Peres 2015; Wickson 2016).

To encourage commercial plant breeding, countries have established exclusive rights tailored to this sector, that is, plant breeders' rights. To qualify for such rights, farmers and breeders need to develop varieties that meet the requirements of novelty, uniformity, distinctiveness and stability (Prifti 2015; De Schutter 2011). In addition, and in contrast with patents, breeders' rights are only awarded for varieties that physically exist (Correa 1995). Plant breeders' rights acknowledge that breeding is a cumulative effort and include a "breeder's exemption" to use under legally specified circumstances protected plant varieties without a license to develop distinctive varieties (Louwaars et al. 2013). On farms, these requirements, and the impermissibility for farmers in many jurisdictions of using protected varieties without a license as raw material for further breeding, exert pressure to reduce diversity (Halpert and Chappell 2017), since this is costly in terms of licensing negotiations and fees.

The commercialization of new plant varieties does not solely rely on intellectual property rights, but also on concentrated efforts to develop hybrid varieties which make the collection of seeds for reseeding unattractive (Halewood 2013). Other research ideas, such as making seeds from crops infertile through Genetic Use Restriction Technology—nicknamed by critics as "terminator seeds"—have provoked public outrage and until now have never been commercialized (Thompson 2009; Glover 2010).

2.3.3 *Copyright*

The idea behind copyright is to protect authentic expressions. To obtain copyright for one's own work it is generally not required to make a formal application. Nowadays copyright affects a large spectrum of areas, ranging from computer software to academic articles. While one remains free to publish similar ideas in distinct forms, current scientific publishing customs make it difficult for others to publish similar ideas in less restrictive commercial outlets. It would even be considered a form of self-plagiarism, and a violation of transferred copyrights, to produce more accessible versions of one's own best work.

Copyright also consists in a bundle of rights. The two broad categories of rights are those protecting the moral interests of authors and those conducive to securing economic interests. Listed under the moral rights of authors, we can find the right to attribution of authorship, which is in principle inalienable, and the right to integrity, whose aim is to protect the interests an author may have in keeping the work intact (Conde Gutiérrez 2011). As mentioned before, in practice authors usually grant editors and publishers some discretion to modify their original expression. Many works are simply labelled as edited, when formal consent cannot be sought. As for economic interests, authors usually have to rely on publishers to commercialize their works, and this often leads to insufficient or no financial returns.

Nowadays there are broad groups who oppose traditional copyright usages. Open science advocates fight for open access to scholarly publications, especially when this is the fruit of research funded with taxpayers' money (Willinsky 2006). We can observe a similar struggle in the fight for open educational resources, to make textbooks and course materials accessible to all (Atenas et al. 2014). These groups are raising awareness of the costs that restrictive copyright policies have for scientific literacy and cooperation. They urge people to consider alternatives to copyright, and in particular to think about the different sub-rights they really want to keep, such as the right to attribution of authorship, the right to the integrity of one's work and the right to non-commercial exploitation. By separating these rights, authors can allow others to engage in socially beneficial and creative activities (Merges 2010), such as redesigning and expanding textbooks, and making translations that are offered on a non-commercial basis. We will return to this point in the next chapter.

2.3.4 *Geographic Indications*

Geographic indications are another form of intellectual property that may protect certain types of innovation, although initially with a significant delay (Raustiala and

Munzer 2007). Regions who have come up with particularly original production methods and product uniqueness can obtain internationally recognized protection for their whole geographic area. Among the most common products that benefit from geographic indications are cheese, olive oil, wine, beer and spirits. Albeit in highly uneven proportions, countries in both the Global North and the Global South have benefited from this form of intellectual property, as examples like Champagne and Tequila illustrate. Due to the complex processes involved in securing geographic dominations, it is usually the already better-off regions that benefit from such exclusive rights.

A major disadvantage, and one might also say injustice, caused by geographic indications is that they only protect innovations from a given geographic area, and thus they leave people who move out of these areas and continue to produce authentic products without protection (Sunder 2007). This makes it extremely costly for artisans to move to other areas, for example when they marry people from other regions, or due to conflicts, natural disasters or changing climatic conditions.

2.3.5 Trademarks

Trademarks may have an indirect effect on agricultural innovation. One of the main justifications for trademarks is that they help to identify a product or service with a specific producer or retailer. This can have substantial effects on agriculture if consumers expect a specific taste or appearance, or a certified form of production, in foodstuffs protected by trademarks. This can have socially welcome effects when a given brand advertises itself as fair-trade, carbon-neutral or generally environmentally friendly. Yet it can have quite disastrous effects for society and the environment if consumers expect uniformity of produce, as this demands large-scale monocultures and discarding for purely aesthetic reasons fruits and vegetables that do not meet the trademark's standards, leading to less resilient food systems and waste. This is a particularly worrying outcome in a world where such a large portion of the population is completely detached from the realities of food production (Thompson 2015; Korthals 2018).

2.3.6 Contracts

Although contracts do not fall directly under intellectual property rights, they do play a special role in proprietary science. Sometimes they give additional support to proprietary rights by obliging users to follow specific indications, as commonly seen in material transfer agreements (MTAs). In the case of seeds, they can forbid farmers to utilize the seeds for the next season or even to follow specific indications on how to use them (Robaey 2016). Material used for plant breeding is also often subject to specific conditions, which may vary from observing international treaties on genetic resources, to avoiding commercial use or sharing sales revenues (Louwaars 2012). Contracts may also prevent scientists from working in other laboratories or publishing certain results, alluding to trade secrets: something that is worrying due to increasingly lax labour laws and reluctance to hire early-career researchers on a long-term basis.

Licencing contracts allow innovation owners to practice corporate social responsibility by including clauses forbidding specific socially undesirable uses, and ensuring through courts that these clauses are respected. For example, the Broad Institute licensed its CRISPR (clustered regularly interspaced short palindromic repeats) patents to Monsanto for agricultural purposes, on the condition that it does not use them to (1) perform gene drives that spread modified genes that may alter ecosystems, (2) create sterile seeds, or (3) develop tobacco products (Guerrini et al. 2017).

2.4 Main Problems and Aims of Justice

When using intellectual property rights as the main innovation incentive scheme in a world where there is extreme inequality in purchasing power, scientific and technological capacities and infrastructure, legal expertise, and institutional strength and stability, a number of problems for social justice tend to emerge. These problems are particularly pressing when delayed access or the unavailability of innovation negatively affect food production. As mentioned before, it is crucial to keep in mind that inventions have an effect that goes far beyond the relation between inventors and their customers, affecting people and their social relations, animals and the environment (Baumgartner 2006; Hongladarom 2015). Climate change is also sharply increasing the need for innovations for adapting to the new climatic conditions, raising additional issues of justice (Robaey and Timmermann 2018; Biddle 2016).

In this section, we have introduced both the ideal functioning of intellectual property regimes, strongly focussing on patents, and the real-world challenges such regimes face in securing social justice, sustainable development and competitive markets. Let us review the effects of these regimes in relation to the six dimensions of social justice introduced in the previous chapter.

The access problem and justice in exchange. The main justification for patent regimes is that they make the fruits of innovation accessible to the public at large. Yet innovation is often costly, and has to be recouped by selling the resulting products to customers with sufficient purchasing power to cover the original research and development costs, as well as ensuring sufficient profits to make research funding commercially attractive. One of the main justifications for scientific advancement and its subvention is that it will ultimately improve social welfare (Derclaye 2012). Yet here there is a major issue for social justice: by asking consumers to pay for the objects of innovation, many people will end up empty-handed. This means that they will be deprived of access to innovations: an outcome that is contrary to the very goal and justification of the intellectual property system, which is to increase the availability of innovations (Baumgartner 2006).

Market incentives may also be detrimental to ecological needs. Short-term economic gains may come at the cost of selling large amounts of a single crop variety, thereby contributing to genetic erosion and reducing the overall ecological resilience of agroecosystems. Similarly, greedy salespersons may persuade farmers to apply much more pesticide than needed to secure more profits.

It is crucial to keep in mind that intellectual property rights are temporary. In the case of patents, after a 20-year period of exclusivity the invention enters the public domain. In theory, this should allow people to access a cheaper generic version of the invention after the patent protection time elapses. In practice, however, this delay comes at the price of not being able to follow eventual commercial standards, of using a less effective pesticide due to pesticide resistance (Timmermann 2015), or of using a plant variety that is more vulnerable to pathogens that could have evolved (Goeschl and Swanson 2003). Timely access to innovation can be crucial for securing the human right to food.

We will return to this issue in the chapter on justice in exchange, and to some of the subsequent problems in the chapter on intergenerational justice.

The availability problem and distributive justice. Intellectual property regimes allow the market to set the highest rewards for those products that find the largest markets. In theory, the benefits of such a system are that there is no need for a central agency that identifies research needs, evaluates innovative capacity, and sets adequate rewards. It would not be cost-efficient to identify such information, and according to economic principles, the market rather than public institutions is best suited to distribute rewards to efficient research enterprises. Moreover, by not relying on a central agency to distribute rewards, corruption and favouritism are avoided (Sterckx 2006).

When applying this reasoning to the real world, we face a major problem: markets may not adequately reflect social and environmental needs. In a world of extreme inequality in purchasing power, economic incentives favour satisfying the wants of wealthy customers instead of the urgent needs of a far larger number of poor people (Hollis and Pogge 2008). We will return to this issue in the chapter on distributive justice.

Facilitating participation and contributive justice. Historically, the publication of the details needed to reproduce an invention in patent documents has facilitated an environment of openness, as the period of exclusivity was not felt as a major constraint, in contrast with trade secrecy. In earlier times, patents and copyright were also awarded for a shorter period of time, and were far harder or even impossible to apply internationally (May 2007). Nowadays, when a large amount of information can be exchanged in seconds, and science is being advanced by groups dispersed around the world, this delay has become disproportionally long. Furthermore, many innovators have learned to extend the actual protection of their invention by engaging in a series of tricks to extend patent life: a practice that has been nicknamed the "ever-greening" of patents (Dwivedi, Hallihosur, and Rangan 2010). Many patents for a single invention also make participation in its improvement difficult. This is particularly problematic when genetic resources are involved, as it is not possible to invent around. Here we should not forget the large discrepancies in purchasing power between different research institutes, and their effect on securing licenses to use the invention.

Exclusive research environments and lack of access to scientific literature have serious problems for global justice, particularly as they hinder possibilities of interacting as peers in scientific projects and technology assessments that are of common interest to all humanity, or at least to people with similar environmental and social conditions. Patents may in certain cases also allow their owners to hinder independent tests and studies to validate claims, and to challenge them in courts (Murray 2006).

We will later use the framework of contributive justice to discuss the social justice problems of an innovation system that obstructs participation.

Encouraging sustainability and intergenerational justice. Linking financial rewards to the diffusion of inventions creates incentives to sell more products than are socially and ecologically prudent, and this may also raise issues of intergenerational justice. In other cases, intellectual property creates artificial scarcity, leading to undersupply, even when an adequate distribution of the goods would be more socially and environmentally desirable. However, when large markets can be expected, due to changing needs or increasing public consciousness, intellectual property may make it profitable to invest large amounts of resources in research. In the case of environmentally friendly technologies, an added advantage is that the private sector also has a strong interest in the diffusion of technologies among paying customers, and therefore actively engages in customer outreach. The temporariness of patents also means that future generations will have more knowledge available in the public domain.

In the chapter on intergenerational justice, we will analyse how innovation affects the realization of the rights of future generations.

Aligning good scientific practice with procedural justice. Innovators who seek patent rights have an incentive to keep their invention secret until they file a patent, in order to meet the novelty requirement. This makes it particularly difficult to engage with stakeholders and provide them the information needed for a critical technology assessment. Deliberations will only truly begin after substantial investments in a research trajectory have been made, and this creates incentives to avoid disclosing negative information. Intellectual property can also be used to limit independent testing, either by not granting licences or by charging prohibitive fees for them.

Arbitration of disputes and restorative justice. While intellectual property allows ample room to recognize various intellectual contributors and share financial rewards, financial incentives may create the temptation to deprive people of their fair share of financial rewards, and not give them credit for their intellectual contributions. Sometimes the mere fear that any of the parties involved will oppose the commercialization of the invention, or ask for unviable financial compensation, may suffice to motivate depriving them of formal recognition.

The obligation to disclose information on the patent document gives third parties the chance to review technologies. It is important, however, that patent holders and courts allow third parties a certain freedom to perform biosafety and product efficacy studies, and that patent holders do not use their exclusive rights to block independent studies. The quality of information in patent documents needs to be improved, and information about where samples came from should be mandatory, to make sure access and benefits principles are respected.

In relation to principles of restorative justice, we will identify and analyse some of the injustices that require arbitration measures.

So as to have a point of comparison concerning the problems and opportunities that result from relying on intellectual property, we will discuss in the next chapter the various social justice implications of non-proprietary science. Thereafter, we will proceed to discuss each of the six social justice dimensions in detail.

References

Anderson, Molly D. 2018. Scientific knowledge of food and agriculture in public institutions: Movement from public to private goods. In *Routledge Handbook of Food as a Commons*, edited by José Luis Vivero-Pol, Tomaso Ferrando, Olivier De Schutter and Ugo Mattei, 185–202. Oxon & NewYork: Routledge.

Atenas, Javiera, Leo Havemann, and Ernesto Priego. 2014. Opening teaching landscapes: The importance of quality assurance in the delivery of open educational resources. *Open Praxis* 6 (1): 29–43.

Attas, Daniel. 2008. Lockean justifications of intellectual property. In *intellectual property and theories of justice*, ed. Axel Gosseries, Alain Marciano, and Alain Strowel, 29–56. Houndmills: Palgrave Macmillan.

Baumgartner, Christoph. 2006. Exclusion by inclusion? On difficulties with regard to an effective ethical assessment of patenting in the field of agricultural bio-technology. *Journal of Agricultural and Environmental Ethics* 19 (6): 521–539.

Beitz, Charles R. 2005. The moral rights of creators of artistic and literary works. *Journal of Political Philosophy* 13 (3): 330–358.

Belleflamme, Paul. 2006. Patents and incentives to innovate: Some theoretical and empirical economic evidence. *Ethical Perspectives: Journal of the European Ethics Network* 13 (2): 267–288.

Bessen, James, Jennifer Ford, and Michael J. Meurer. 2011. The private and social costs of patent trolls. *Regulation* 34: 26–35.

Biagioli, Mario. 2006. Patent republic: Representing inventions, constructing rights and authors. *Social Research: An International Quarterly* 73 (4): 1129–1172.

Biddle, Justin B. 2014. Can patents prohibit research? On the social epistemology of patenting and licensing in science. *Studies in History and Philosophy of Science Part A* 45: 14–23.

Biddle, Justin B. 2016. Intellectual property rights and global climate change: Toward resolving an apparent dilemma. *Ethics, Policy & Environment* 19 (3): 301–319.

Breakey, Hugh. 2009. Liberalism and intellectual property rights. *Politics, Philosophy & Economics* 8 (3): 329–349.

Chopra, Samir. 2018. End intellectual property. Aeon, Accessed March 14, 2019. https://aeon.co/essays/the-idea-of-intellectual-property-is-nonsensical-and-pernicious.

Conde Gutiérrez, Carlos. 2011. Copyrights y derechos morales de autor: la experiencia del common law en el Reino Unido. *Revista La Propiedad Inmaterial* 15 (November).19–29.

Conde Gutiérrez, Carlos. 2017. Invenciones Contrarias al Orden Público y Moral. In *Derecho de Patentes*, edited by Ernesto Rengifo García, 503–534. Bogotá: Universidad del Externado.

Correa, Carlos M. 1995. Sovereign and property rights over plant genetic resources. *Agriculture and Human Values* 12 (4): 58–79.

Cwik, Bryan. 2014. Labor as the basis for intellectual property rights. *Ethical Theory and Moral Practice* 17 (4): 681–695.

De Jonge, Bram. 2011. What is fair and equitable benefit-sharing? *Journal of Agricultural and Environmental Ethics* 24 (2): 127–146.

De Schutter, Olivier. 2011. The right of everyone to enjoy the benefits from scientific progress and the right to food: from conflict to complementarity. *Human Rights Quarterly* 33: 304–350.

Derclaye, Estelle. 2012. Eudemonic intellectual property: Patents and related rights as engines of happiness, peace, and sustainability. *Vanderbilt Journal of Entertainment and Technology Law* 14 (3): 495–543.

Drahos, Peter. 1996. *A philosophy of intellectual property*. Aldershot: Dartmouth.

Drahos, Peter, and John Braithwaite. 2003. *Information feudalism: Who owns the knowledge economy?*. New York: New Press.

Dwivedi, Gaurav, Sharanabasava Hallihosur, and Latha Rangan. 2010. Evergreening: A deceptive device in patent rights. *Technology in Society* 32 (4): 324–330.

Feeney, Oliver, Julian Cockbain, Michael Morrison, Lisa Diependaele, Kristof Van Assche, and Sigrid Sterckx. 2018. Patenting foundational technologies: Lessons from CRISPR and other core biotechnologies. *The American Journal of Bioethics* 18 (12): 36–48.

Forsberg, Ellen-Marie, Anders Braarud Hanssen, Hanne Marie Nielsen, and Ingrid Olesen. 2017. Patent ethics: The misalignment of views between the patent system and the wider society. *Science and Engineering Ethics*, 1–26.

García Márquez, Gabriel. 2002. *Vivir para contarla*. Bogotá: Diana.

Glover, Dominic. 2010. The corporate shaping of GM crops as a technology for the poor. *The Journal of Peasant Studies* 37 (1): 67–90.

Goeschl, Timo, and Timothy Swanson. 2003. Pests, plagues, and patents. *Journal of the European Economic Association* 1 (2–3): 561–575.

Gosseries, Axel. 2011. Piratear es robar? *Dixit* 14: 37–39.

Guerrini, Christi J., Margaret A. Curnutte, Jacob S. Sherkow, and Christopher T. Scott. 2017. The rise of the ethical license. *Nature Biotechnology* 35 (1): 22–24.

Halewood, Michael. 2013. What kind of goods are plant genetic resources for food and agriculture? Towards the identification and development of a new global commons. *International Journal of the Commons* 7 (2): 278–312.

Halpert, Madeleine-Thérèse, and M Jahi Chappell. 2017. Prima facie reasons to question enclosed intellectual property regimes and favor open-source regimes for germplasm. *F1000Research* 6:284.

Hassoun, Nicole. 2015. The global health impact index: Promoting global health. *PLoS ONE* 10 (12): e0141374.

Haugen, Hans Morten, Manuel Ruiz Muller, and Savita Mullapudi Narasimhan. 2011. Food security and intellectual property rights: finding the linkages. In *Intellectual property and human development: Current trends and future scenarios*, ed. Tzen Wong and Graham Dutfield, 103–138. Cambridge: Cambridge University Press.

Heins, Volker. 2008. Human rights, intellectual property, and struggles for recognition. *Human Rights Review* 9 (2): 213–232.

Hettinger, Edwin C. 1989. Justifying intellectual property. *Philosophy & Public Affairs* 18 (1): 31–52.

Hollis, Aidan, and Thomas W. Pogge. 2008. The health impact fund: Making new medicines accessible for all. In Oslo & New Haven: Incentives for Global Health. http://www.healthimpactfund.org/hif_book.pdf. Accessed 31 March 2012.

Hongladarom, Soraj. 2015. Intellectual property rights and food security: The role of external relations. In *Food Security and Food Safety for the Twenty-first Century*, 255–262. Singapore: Springer.

Hsieh, Nien-hê. 2000. Moral desert, fairness and legitimate expectations in the market. *Journal of Political Philosophy* 8 (1): 91–114.

Hughes, Justin. 1988. The philosophy of intellectual property. *The Georgetown Law Review* 77: 287–366.
Hull, Gordon. 2009. Clearing the rubbish: Locke, the waste proviso, and the moral justification of intellectual property. *Public Affairs Quarterly* 23 (1): 67–93.
Jacob, Claus, and Adam Walters. 2005. Risk and responsibility in chemical research: The case of agent orange. *HYLE–International Journal for Philosophy of Chemistry* 11 (2): 147–166.
James, Aaron. 2012. *Fairness in practice: A social contract for a global economy*. New York: Oxford University Press.
Jefferson, Thomas. 1813. Letter to Isaac McPherson—August 13, 1813. In Wikisource. https://en.wikisource.org/wiki/Letter_to_Isaac_McPherson_-_August_13_1813.
Kapczynski, Amy. 2008. The access to knowledge mobilization and the new politics of intellectual property. *Yale Law Journal* 117 (5): 804–885.
Kloppenburg, Jack. 2005. *First the seed: The political economy of plant biotechnology*, 2nd ed. Madison: University of Wisconsin Press.
Korthals, Michiel. 2018. Is intensive farming ethically acceptable? *Annals of Advanced Agricultural Sciences* 2 (2): 15–29.
Korthals, Michiel, and Henk van den Belt. 2014. Intellectual property rights and trade in the food and agricultural sectors. In *Encyclopedia of Food and Agricultural Ethics*, edited by Paul B. Thompson and David M. Kaplan, 1271–1278. Dordrecht: Springer Netherlands.
Lambrecht, Maxime. 2017. The time limit on copyright: An unlikely tragedy of the intellectual commons. *European Journal of Law and Economics* 43 (3): 475–494.
Landes, William M., and Richard A. Posner. 2004. *The political economy of intellectual property law*. Washington: American Enterprise Institute.
Lewin, Peter. 2007. Creativity or coercion: Alternative perspectives on rights to intellectual property. *Journal of Business Ethics* 71 (4): 441–455.
Liivak, Oscar, and Eduardo M. Peñalver. 2013. The right not to use in property and patent law. *Cornell Law Review* 98: 1437–1493.
Locke, John. 1689. *Two Treatises of Government*. Edited by Peter Laslett. Cambridge: Cambridge University Press.
Louwaars, Niels. 2012. Seed science in the 21st century: rights that scientists have to deal with. *Seed Science Research* 22 (S1): S9–S14.
Louwaars, Niels, Bram De Jonge, and Peter Munyi. 2013. "Intellectual Property Rights in the Plant Sciences and Development Goals in Agriculture: An Historical Perspective." In *Knowledge Management and Intellectual Property*, edited by Stathis Arapostathis and Graham Dutfield, 252–272. Cheltenham & Northampton: Edward Elgar.
Love, James, and Tim Hubbard. 2007. The big idea: Prizes to stimulate R&D for new medicines. *Chicago-Kent Law Review* 82 (3): 1519–1554.
Mancilla, Alejandra. 2012. Noncivil disobedience and the right of necessity. A point of convergence. *Krisis* 3: 3–16.
May, Christopher. 2007. The hypocrisy of forgetfulness: the contemporary significance of early innovations in intellectual property. *Review of International Political Economy* 14 (1): 1–25.
Merges, Robert P. 1994. Of property rules, coase, and intellectual property. *Columbia Law Review* 94 (8): 2655–2673.
Merges, Robert P. 2010. Individual creators in the cultural commons. *Cornell Law Review* 95: 793–805.
Merges, Robert P. 2011. *Justifying intellectual property*. Cambridge, MA: Harvard University Press.
Moore, Adam. 2008. Personality-based, rule-utilitarian, and lockean justifications of intellectual property. In *The handbook of information and computer ethics*, edited by Kenneth Einar Himma and Herman T. Tavani, 105–130. Hoboken: Wiley.
Moore, Adam. 2011. Intellectual Property. In *The Stanford Encyclopedia of Philosophy (Spring 2011 Edition)*, edited by Edward N. Zalta.
Moore, Jason W. 2018. The Capitalocene Part II: accumulation by appropriation and the centrality of unpaid work/energy. *The Journal of Peasant Studies* 45 (2): 237–279.

Murray, Kali N. 2006. Rules for radicals: A politics of patent law. *Journal of Intellectual Property Law* 14: 63–110.

Nozick, Robert. 1974. *Anarchy, state, and Utopia*. New York: Basic Books.

O'Neill, John. 2018. Food, needs and commons. In *Routledge Handbook of Food as a Commons*, edited by José Luis Vivero-Pol, Tomasso Ferrando, Olivier De Schutter and Ugo Mattei, 103–120. Oxon & New York: Routledge.

Papaioannou, Theodoros. 2006. Towards a critique of the moral foundations of intellectual property rights. *Journal of Global Ethics* 2 (1): 67–90.

Peres, Sara. 2015. Saving the gene pool for the future: Seed banks as archives. *Studies in History and Philosophy of Science Part C: Studies in History and Philosophy of Biological and Biomedical Sciences*. https://doi.org/10.1016/j.shpsc.2015.05.002.

Pogge, Thomas W. 2010. The Health Impact Fund: better pharmaceutical innovations at much lower prices. In *Incentives for global health: Patent law and access to essential medicines*, ed. Thomas W. Pogge, Matthew Rimmer, and Kim Rubenstein, 135–154. Cambridge: Cambridge University Press.

Prifti, Viola. 2015. *The breeder's exception to patent rights: Analysis of compliance with Article 30 of the TRIPS agreement*. Cham et al.: Springer.

Prifti, Viola. 2016. Die Rolle des öffentlichen Interesses an Pflanzenpatenten: Eine europäische Perspektive. In *Biopatente*, ed. Barbara Brandl and Stephan Schleissing, 55–74. Baden-Baden: Nomos.

Radder, Hans. 2010. The commodification of academic research. In *Science and the modern university*, ed. Hans Radder, 1–23. Pittsburg: University of Pittsburg Press.

Radder, Hans. 2013. Exploring Philosophical Issues in the Patenting of Scientific and Technological Inventions. *Philosophy & Technology* 26 (3): 283–300.

Raustiala, Kal, and Stephen R. Munzer. 2007. The global struggle over geographic indications. *European Journal of International Law* 18 (2): 337–365.

Resnik, David B. 2003. A pluralistic account of intellectual property. *Journal of Business Ethics* 46 (4): 319–335.

Resnik, David B. 2016. Scientific realism and the patent system. *Journal for General Philosophy of Science* 47 (1): 69–77.

Robaey, Zoë. 2016. Transferring moral responsibility for technological hazards: the case of GMOs in agriculture. *Journal of Agricultural and Environmental Ethics* 29 (5): 767–786.

Robaey, Zoë, and Cristian Timmermann. 2018. "Fair agricultural innovation for a changing climate." In *Food Justice, the Environment, and Climate Change*, edited by Erinn Cunniff Gilson and Sarah Kenehan, 213–230. Lanham: Rowman & Littlefield.

Schiebinger, Londa. 2004. Feminist history of colonial science. *Hypatia* 19 (1): 233–254.

Schneider, Ingrid. 2010. *Das Europäische Patentsystem. Wandel von Governance durch Parlamente und Zivilgesellschaft*. Frankfurt am Main & New York: Campus.

Sherman, Brad. 2015. What does It mean to invent nature? *UC Irvine Law Review* 5: 1193–1229.

Stephan, Paula E. 2012. *How economics shapes science*. Cambridge, MA: Harvard University Press.

Sterckx, Sigrid. 2006. The moral justifiability of patents. *Ethical Perspectives* 13 (2): 249–265.

Sterckx, Sigrid. 2011. Patenting and licensing of university research: promoting innovation or undermining academic values? *Science and Engineering Ethics* 17 (1): 45–64.

Stiglitz, Joseph E. 1999. "Knowledge as a global public good." In *Global public goods: international cooperation in the 21st century*, edited by Inge Kaul, Isabelle Grunberg and Marc A. Stern, 308–325. New York & Oxford: Oxford University Press.

Strauss, Leo. 1952. On Locke's doctrine of natural right. *The Philosophical Review* 61 (4): 475–502.

Sunder, Madhavi. 2007. The invention of traditional knowledge. *Law and Contemporary Problems* 70: 97–124.

Thompson, Paul B. 2009. Philosophy of Agricultural Technology. In *Philosophy of technology and engineering sciences*, ed. Anthonie Meijers, 1257–1273. Amsterdam: Elsevier.

Thompson, Paul B. 2015. *From field to fork: Food ethics for everyone*. New York: Oxford University Press.

Timmermann, Cristian. 2014a. An assessment of prominent proposals to amend intellectual property regimes using a human rights framework. *La Propiedad Inmaterial* 18: 221–253.

Timmermann, Cristian. 2014b. Limiting and facilitating access to innovations in medicine and agriculture: A brief exposition of the ethical arguments. *Life Sciences, Society and Policy* 10: 8.

Timmermann, Cristian. 2015. Pesticides and the patent bargain. *Journal of Agricultural and Environmental Ethics* 28 (1): 1–19. https://doi.org/10.1007/s10806-014-9515-x.

Timmermann, Cristian. 2017. Harvesting the uncollected fruits of other people´s intellectual labour. *Acta bioethica* 23 (2): 259–269. https://doi.org/10.4067/s1726-569x2017000200259.

Trerise, Jonathan. 2010. Against the strength of patent protection. *The Monist* 93 (3): 464–480.

Trerise, Jonathan. 2016. The influence of patents on science. *Politics, Philosophy & Economics* 15 (4): 424–450.

Tvedt, Morten Walløe, and Ellen-Marie Forsberg. 2017. The room for ethical considerations in patent law applied to biotechnology. *The Journal of World Intellectual Property* 20 (5–6): 160–177.

Van Overwalle, Geertrui. 2009. *Patents in agricultural biotechnology and the right to food*. Geneva: United Nations.

Van Overwalle, Geertrui. 2010. Turning patent swords into shares. *Science* 330 (6011): 1630–1631.

Varelius, Jukka. 2014. Do patents and copyrights give their holders excessive control over the material property of others? *Ethics and Information Technology* 16 (4): 299–305.

Waldron, Jeremy. 1992. From authors to copiers: Individual rights and social values in intellectual property. *Chicago-Kent Law Review* 68: 841–887.

Wickson, Fern. 2016. Do we care about Synbiodiversity? Questions arising from an investigation into whether there are GM crops in the Svalbard global seed vault. *Journal of Agricultural and Environmental Ethics* 29 (5): 787–811.

Widerquist, Karl. 2010. Lockean theories of property: Justifications for unilateral appropiation. *Public Reason* 2 (1): 3–26.

Willinsky, John. 2006. *The access principle: The case for open access to research and scholarship, Digital libraries and electronic publishing*. Cambridge, Mass.: MIT Press.

Wilson, James. 2007. GM crops: Patently wrong? *Journal of Agricultural and Environmental Ethics* 20 (3): 261–283.

Wilson, James. 2009. Could there be a right to own intellectual property? *Law and Philosophy* 28 (4): 393–427.

Wilson, James. 2010. Ontology and the regulation of intellectual property. *The Monist* 93 (3): 450–463.

Chapter 3
Agriculture and Non-proprietary Science

Abstract This chapter provides an overview of the different innovation types and streams that are not covered or insufficiently incentivized by intellectual property regimes. It starts by discussing the benefits of non-proprietary science systems, in relation to increased participation, good scientific practice, the common heritage of humankind, competitive markets and public trust in science. After that it discusses types of innovation that are insufficiently incentivized. It concludes by discussing two agricultural innovation streams that are currently in danger due to the enclosure, expansion and overpromotion of intellectual property rights: traditional ecological knowledge and open science.

Keywords Open innovation · Traditional knowledge · Common heritage of humankind · Participation · Parallel innovation streams

3.1 Introduction

We can distinguish three basic types of invention that find themselves outside the intellectual property incentive system: (i) those inventions that do not meet the formal criteria to qualify for any form of intellectual property protection, (ii) those inventions that meet the legal requirements for intellectual property rights, but are seldom developed and protected because they can easily be reproduced or obtained without paying royalties, and lastly (iii) inventions that are kept openly accessible as a matter of choice or principle. These types of inventions and ways of innovating are insufficiently incentivized, or even under threat, in a world dominated by intellectual property; therefore they must be self-supported or else rely on government funding, prizes or open innovation alliances and licensing schemes.

For this reason, the diminishing funding destined for public institutions is having a particularly detrimental effect on the development of these types of innovation (Tittonell 2013). Moreover, universities and public institutions are increasingly obliged to form public-private partnerships, especially in the form of university-industry alliances, in order to secure the additional funding on which they are increasingly dependent for their effective functioning (Stephan 2012; Sterckx 2011; May

C. Timmermann, *Social Justice and Agricultural Innovation*, The International Library of Environmental, Agricultural and Food Ethics 31,
https://doi.org/10.1007/978-3-030-56193-2_3

2006). As these alliances seek to generate marketable products, that is, private goods as opposed to public goods, they tend to favour research trajectories that produce proprietary innovations. Furthermore, innovation that can easily be replicated is often diffused on a non-commercial basis, and therefore becomes invisible for many development progress indicators, and often even uninteresting for policies designed to increase the national gross domestic product. This has the consequence that innovations or solutions that are easily replicable and do not involve market commodities are often not even considered as part of the solution to global challenges such as fighting hunger and malnutrition, climate change mitigation, biodiversity conservation and soil restoration, particularly when they are targeted at people with little or no purchasing power (Gupta 2010). In general terms, neoliberal government policies are characterized by a strong mandate to strengthen industry, often thereby sacrificing non-proprietary innovation streams. This has major effects on social justice, as private research responds to market forces and shareholders' interests (Piesse and Thirtle 2010), even when these are not in line with social and ecological needs.

Intellectual property not only affects the type of research yielded by an innovation system, but also promotes a certain way of conducting research. For example, the ranking of universities and metrics for academic performance tend to disfavour incremental research, as this research tends to require many years of studies in order to deliver publishable findings (Geertsema et al. 2016; Vanloqueren and Baret 2009). This discourages the study and development of incremental innovation (which is more inclusive) and encourages the pursuit of breakthrough research (which tends to be elitist). It also fosters an ambience of secrecy and a reluctance to share. Commercialization makes competition for funding, industry partners and reputation particularly lucrative. In addition, sharing can jeopardize the novelty requirement of patentability, and is often seen as a lost business opportunity when not leading to profitable material transfer agreements or data use licenses.

In light of the above, the aim of this chapter is to focus on two specific questions. First, what types of innovation do we miss out on by overly relying on intellectual property? Second, what forms of doing research do we fail to promote by overly relying on intellectual property?

I will start by discussing some of the arguments in favour of non-proprietary science, and then highlight the different varieties of research that will remain under-incentivized: basic research, innovation based on principles, innovation that is considered obvious by those skilled in the state of the art, secondary usages of expired patented innovations, and easily reproducible inventions. After that, I will continue to discuss two types of research that are increasingly difficult to carry out: open science and the development of traditional ecological knowledge. Finally, I will examine some of the advantages and disadvantages of keeping these innovation systems working effectively side by side, including the improvement of competition (by providing a freely available standard for comparison), allowing anonymous contributions, establishing alternative distribution channels, and facilitating inclusion and epistemic diversity.

3.2 Arguments in Favour of Non-proprietary Science

A major problem that unites the different groups that object to the overcommercialization of science is that there are different ways to conceptualize innovation outside the intellectual property domain. In an extreme case, one can understand part of science as a *res nullius*, something that is free to be used however one likes, without feeling obliged to provide something in return. Another option is to consider scientific knowledge as being part of the *public domain*. Knowledge under this regime cannot be appropriated and everyone has access to it, irrespective of whether they have contributed or show willingness to contribute to it. The fact that knowledge in the public domain is freely accessible does not mean that contributors are not owed acknowledgment for their authorship (Boyle 2008). Finally, a third option is to understand this knowledge as a *commons* and make it conditional on specified use regulations, such as a commitment to contribute back whatever improvements or added knowledge one gathers by using this commons, and often being subject to community decision-making mechanisms (Hess and Ostrom 2007).

Irrespective of how people tend to conceptualize the place of science in society, we can also observe a number of different arguments that supporters of non-proprietary science use to justify their position. The first line of argument defends participation; the second condemns the commercialization of science as distorting the quest for truth and well-ordered science; the third condemns the enclosure of science by appealing to its status as the common heritage of humankind; the fourth welcomes competition by the public sector; and the fifth focuses on the difficulties in maintaining public trust in science.

3.2.1 Facilitating Participation in Creative Efforts

An ability to participate in science is sometimes demanded as a matter of justice, or even as a human right, similarly to what can be observed with other cultural enterprises (Timmermann 2014; Shaver 2010). Here inclusion is seen as a social right. Thus, as a norm, any scientific project should aim to be inclusive, since being arbitrarily hindered from participating in socially relevant enterprises is seen as an injustice.

In other cases, participation is defended as crucial for speeding up scientific advancement; so access becomes a matter of justice also due to the welfare-enhancing character of science. Expanding participation is here also seen as a matter of justice, not so much for the sake of participation in itself, but rather due to the strong potential that inclusive science has to improve the adequacy of innovation. Open science projects can harvest the perspectives of many more people, allowing a wider set of contributions and identified problems, as well as furthering the identification of needs, thus increasing efficiency in science output and improving the likelihood that science is directed towards solving urgent needs. Participation thus becomes a means to improve social welfare and gain more control over one's circumstances (cf. Steinvorth 2009).

Another argument specifically highlights the benefits of openness as vital for exercising and improving problem-solving capacities. In this sense, open-source software advocates have stated in relation to problem-solving: "Given enough eyeballs, all bugs are shallow" (Raymond 1999, 29). Closed science projects do not allow indirect verification or user innovation, thus hindering follow-up innovation and the development of add-ons (Torrance and von Hippel 2015). People should not be impeded from exercising their ability to develop and refine tools, adapting them to their specific needs and wants: something that characterizes humans as a species. The central driver for participation, from this perspective, is to allow people to satisfy their intellectual curiosity and build tools.

Participation has both intrinsic and instrumental value for people. A full defence of this position will be offered in Chap. 6, in relation to the idea of contributive justice.

3.2.2 Good Scientific Practice

One of the main criticisms of proprietary science models, particularly from philosophers of science, is that they jeopardize good scientific practice. A common approach to highlighting the weaknesses of this system is to refer to Robert Merton's key scientific imperatives: universalism, communalism, disinterestedness, and organized scepticism (Merton 1942; van den Belt 2010; Macfarlane and Cheng 2008; Koepsell 2010). Here is a short overview of Merton's imperatives from today's perspective.

Universalism. The first imperative identified by Merton is universalism. Scientific validation should not depend on the personal or social attributes of key scientific figures and their testimony. Here the question of who developed a theory or made an observation becomes scientifically irrelevant. All prejudices and forms of discrimination have to be set aside. Science has an impersonal character. Therefore nationalistic biases and prejudices, as exemplified by the condemnation of "enemy science", need to be left aside. Here it is important to note that these norms were written during World War II, testifying to huge divisions in the scientific community due to war, racism, and differences in political and economic ideology.

Another demand of universalism is that scientific careers and scientific projects be open to talent. Similarly, as with ignoring ideas because of their origin, it is detrimental for science (and society) to block the entry of people who come from outside one's community or class. Therefore, measures need to be taken to preserve and establish equality of opportunity for participants all over the world.

Communalism. The norm of communalism, formerly known as "communism", highlights the importance of research findings being part of the public domain. This is for two reasons. First, as the substantive findings of science originate from scientific collaboration, it is inaccurate and illegitimate to attribute them to single individuals.

Scientists may claim recognition and esteem for their efforts, but not ownership of the findings they claim as "theirs", since these findings would be unconceivable without the contributions of others and the prior work they build upon.

Second, scientific findings are building blocks and tools that allow us to continue with the advancement of science. Common ownership of research results is crucial for promoting collective collaboration. This requires collaboration between past and present generations, to safeguard our common scientific heritage so that scientific knowledge can be expanded and revised. Here we also need to consider the global dimension of scientific research. In a world of extreme inequality, today's already exorbitant fees for scientific publications, as well as some types of equipment and materials, are impediments to integrating people from all over the world into scientific efforts.

Common ownership also involves an obligation to fully and openly communicate one's findings. Sharing the "wealth of science" becomes morally compulsory. Scientific findings are meant to be disclosed and not archived in personal notebooks. This condemns trade secrecy, delayed publication, and even excessive modesty in publishing results.

Disinterestedness. Financial stakes in innovation may bias organizations and individuals towards hiding the eventual harms of an invention, and may even motivate making up errors in competitors' works (Stephan 2012; Fernández Pinto 2015). Profit-driven science may distort scientific decision-making, by affecting the framing of research questions, the selection and interpretation of scientific evidence, the methodologies used, and the reporting of results (Intemann and de Melo-Martín 2014). Scientists should act with the aim of advancing science, and not be strongly influenced by interests of a merely personal nature. This requires that scientists are not too attached to their ideas, as this may distort their critical analysis. It also requires one to be critical of the current and historical contributions of the scientific institutions one is personally or professionally attached to, avoiding any biases in favour of one's alma mater or the achievements of one's compatriots. The only interest scientists should pursue is the quest for truth. Scientific findings need to be constantly checked, and scientists should welcome and encourage revisions to their work.

Another factor distorting science is discrimination. Feminist social epistemologists condemn the effects of disregarding feminist perspectives (Intemann and de Melo-Martín 2014), which take into account issues such as the intellectual weight given to women's contributions, and the effects innovations have on women, as well as other factors that have been amply examined by feminist scholars, such as vulnerability, the recognition of care needs, and changes between people's life stages (Plumwood 2006; Gilson 2015; Preston and Wickson 2016). A similar charge is made by scholars working on decolonization, who accuse formal scientific research of largely ignoring contributions to science from the Global South, particularly the perspectives and innovations of indigenous peoples (Grey and Patel 2015; Kelbessa 2015).

It is debatable to what extent, if at all, the principle of disinterestedness is compatible with efforts to give more space to, or systematically integrate into common scientific systems, those perspectives that have suffered historical discrimination and have been neglected mainly due to present and past power structures. Since this principle is primarily forward-looking, once an ethos for good scientific practice is institutionalized, it does not systematically address issues of past bias or how to address historical wrongdoings. We will return to this point in Chap. 9, on restorative justice.

Organized scepticism. Finally, openness facilitates the scientific scrutiny of methods and observations. Science needs to be sceptical about how results were obtained. This requires experiments to be adequately described so as to allow replication, under the same or different conditions. For example, the trend of writing patent documents in such a way that they do not allow replication of the invention is contrary to this principle. Organized scepticism would also require industry to lose the liberty to include clauses in collaborations with universities and public institutions whereby they are granted a right to delay publications or prohibit information from being made public (di Norcia 2005). Nothing should be considered unquestionable or beyond scientific scrutiny. There are no ideas in science that should not be questioned or objectively analysed. This means that all knowledge, regardless of whether it comes from formal science or traditional knowledge systems, or has special cultural or religious significance, needs to be verified and questioned for it to be considered scientific. An instrumental treatment of knowledge may lead to conflicts with the customary laws of indigenous peoples (Taubman 2005), the resolution of which would require special public education programmes to help the affected people understand the need for a sceptical approach in knowledge acquisition and evaluation.

3.2.3 Common Heritage

Another argument against proprietary science involves characterizing science as the common heritage of humankind. While the basic idea of common heritage has deep historic roots, the concept of a common heritage of humankind—referring to goods from which all humankind should benefit, whether in the present or in the future—emerged in the 1960s as a result of envisioning the political and legal status of the deep sea and outer space (Wolfrum 1983). The principle of common heritage involves (1) common management, (2) no unilateral appropriation without the worldwide sharing of benefits, (3) the rapid sharing of knowledge obtained from scientific research, (4) the prohibition of harmful uses, and (5) preservation for future generations (Timmermann and Robaey 2016; Shackelford 2008). The principle invokes both rights and obligations. Thus states would not only have the right to access the fruits of scientific research, but would also be obliged to contribute to the conservation of scientific knowledge: something that, if taken seriously, would require substantial resources, as does the conservation of traditional knowledge and cultural goods. In agriculture this is a major issue, since not only knowledge has to be conserved, but also the associated genetic resources (Peres 2016). Infrastructure like that of the Svalbard Global Seed Vault in the Arctic, sometimes referred to as the "doomsday seed bank", illustrate the massive effort involved in such undertakings (Wickson 2016).

Most scientists and many members of society see science as something that improves social welfare. This is one reason why many constitutions and human rights declarations promote scientific advancement and demand that the fruits of research be widely shared (Besson 2015). When science is appropriated by a small group to increase rent-seeking, many will not perceive scientific advancement as something intrinsically good, especially if people are thereby exploited and do not benefit from increased automation, or even worse end up losing their food entitlement. To avoid such outcomes, research agendas need to be set in accordance with public welfare, and special programmes need to be implemented to avoid negative outcomes and institute a fairer distribution of costs and benefits, in order to prevent situations, for example, where technological advancement leads to massive job losses without any form of compensation. A common-heritage understanding of scientific knowledge highlights the importance of fair and rapid knowledge sharing.

A crucial problem with understanding scientific knowledge as common heritage is that currently people contribute to this pool at very different rates, with some making much smaller contributions than others, even though they have the means to engage in much more scientific research. In spirit, the principle of the common heritage of humankind calls for a reciprocity-based system: if we want to understand scientific knowledge as common heritage, countries need to reach a consensus on what would be a fair contribution to science, for example, a given percentage of their GDP according to their rate of development. Countries in a similar development stage that are below this rate need to increase their contributions in order to avoid free-riding on the efforts of other countries (Timmermann and van den Belt 2012). We are very far away from this goal, considering research investments in countries like Chile and Romania, who according to the World Bank data sets spent far less resources on science than average in relation to their GDP.[1]

3.2.4 Competition and Breaking Private Monopolies

Non-proprietary science has the added benefit of offering a freely available product of a certain standard, on top of which for-profit science needs to provide sufficient added benefits, in terms of improved quality or desirable complements, in order to be worth the extra money. Therefore public science can have a monopoly-breaking effect, as the private sector will have to offer additional benefits to justify the sales price of competing products (Eisenberg and Nelson 2002). This adds welcome competition that benefits the public by helping to break up usurious monopolies. This allows society at large to benefit from scientific advancement, and may also facilitate the further development of research that was halted because of prohibitive licencing conditions (i.e. so-called “locked-in” research). The ineffective way in which governments have too often dealt with monopolies, and the failure to establish and enforce consumer-friendly anti-trust laws, make the availability of public

[1]See research and development expenditure (% of GDP) at https://data.worldbank.org/.

versions highly necessary. This is so not only for consumer end-products, but also for research tools and some genetic resources, which may even include key starting materials for experiments (Jefferson 2006).

There have been a few attempts to offer freely available tools and inputs to agriculture. These initiatives often enjoy substantial assistance from universities, individuals, NGOs and foundations. Here we have, for example, the Open Source Seed Initiative (Luby et al. 2016), which offers seeds using creative commons contracts, to facilitate breeding, exchange and reseeding. There have also been efforts to increase the number of research tools for agriculture that are available on a commons basis (Jefferson 2006). We will come back to such initiatives later in this chapter, in a section on open science.

3.2.5 Corporate Ties and Public Trust in Science

The commodification of research at academic and public institutions normalizes ties with private sponsorship. By taking money from the private sector, especially from major corporations, public institutions lose their independence, and from the perspective of large segments of the public, also the impartiality of their perspectives and findings. Corporate funding for science projects and salary supplements may lead to significant conflicts of interests. These institutions' lack of independence from their corporate ties triggers an overall distrust in science when corporations are involved in scandals (Radder 2010). This is particularly problematic for a society that relies so intensively on technology, and that now urgently needs to tackle global collective action problems such as climate change. Without public trust in science and adequate scientific education campaigns, alarms and warnings from the scientific community can fall on deaf ears, as is currently seen with anti-vaccination movements (Goldenberg 2016).

3.3 Types of Under-Incentivized Research

Let us briefly review the types of innovation and research approaches we miss out on by overly relying on intellectual property incentives. The main problem for this kind of research is that it has a public-good character. This means that it is not possible to cost-effectively exclude people that did not contribute to its development and yet benefit from its existence. Second, commercial actors in particular would benefit from the existence of such research, but they are often unable to individually dedicate enough resources to its production, and at the same time maintain economic competitiveness, while others continuously free-ride on their efforts. The result is that such research is often not done, unless the different institutions organize themselves and pool their resources, for example via consortia or public-private partnerships, as is seen in many large-scale genome sequencing efforts.

3.3.1 *Basic Research*

The type of research that is most often described as insufficiently funded by proprietary science is what is called "basic science" or "foundational research" (Stiglitz 1999). This outcome is not only condemned by universities, but also by industry. Even major corporations who generally benefit from the patent system recognize that basic research is a public good that is inadequately provided by the market, and therefore requires direct government funding (Trerise 2016). As a public good, basic research faces the collective action problem mentioned above. Many parties would benefit from the advancement of basic science, but individual institutions gain substantially when they free-ride on other institutions' efforts to advance basic research. As a consequence, few institutions find it worthwhile to engage in basic research, as they may lose resources that could be used to gain a competitive advantage: an outcome that is to the detriment of all, as future research is often dependent on solid knowledge of the fundamentals of science. It must be noted, however, that particularly during the post-war decades in the United States, private companies made massive investments in basic research, under the conviction that such investments would give them a competitive advantage in bringing completely new products into the market (Lucier 2019). Hence, once a critical mass invests in basic research on a voluntary basis, the competitive disadvantage of assuming such costs becomes increasingly smaller, allowing the private sector to contribute to such efforts.

3.3.2 *Innovation Based on Principles*

There are large fields of research that generally work without patents, mathematics being the best-known example. Some scientific disciplines have subfields that work with patents, such as agricultural biotechnology, and other subfields, such as agroecology, that work without patents. Research on science and methods that rely on principles that, by law, are not patentable, receive insufficient funding.

As far as agriculture is concerned, agroecologists in particular condemn this trend. Principles act as guidelines for skilled people to establish and improve systems. In agroecology, examples of such principles are the recycling of biomass and the closing of nutrient cycles, thereby assuring favourable soil conditions, and promoting functional biodiversity through smart plant associations, which improves biological interactions and synergies (Altieri 2002; Gliessman 2007). Farming that is based on principles rather than strict recipes demands a stronger skillset from farmers, as they will have to adapt these principles to local environmental and social conditions and locally available resources (Timmermann and Félix 2015). Consequently, innovation based on principles encourages and even demands user innovation (Šūmane et al. 2018). Furthermore, these skills and knowledge can be used to handle other social and environmental issues.

Providing overly high incentives for proprietary science may also distort the approaches taken to a problem. As mentioned in the last chapter, it has been noted in the case of medicine that profit-driven science incentivizes curative approaches much more than it incentivizes public health measures that are preventive in nature (Intemann and de Melo-Martín 2014). The former benefit mostly from patents, while the latter measures are often not patentable, as they involve principles or are considered obvious under patent law. If we translate such reasoning to agriculture, we may expect a similar bias in favour of pesticides and fertilizers over integrated pest management and smart plant associations.

3.3.3 Innovation that Is Considered Obvious by Those Skilled in the Art

Innovation that does not involve a sufficiently large "inventive" step also fails to qualify for patents, as it does not meet one of the formal requirements for patentability, which—in legal terminology—is that inventions are non-obvious to someone skilled in the art or involve an inventive step.[2] Or, as is the case with plant breeders' rights, that they are not stable or different enough from other registered varieties. Such types of innovations are often referred to as "incremental innovation". For example, the adaptation of a composting method to a local environment and the availability of specific residues will often not involve a sufficiently major step to qualify for intellectual property rights, but may nevertheless involve a substantial effort in terms of trial and error, demanding lengthy monitoring, the exchange of experiences, and using up valuable resources that could have been safely spent on proven methods. This type of innovation clearly has enormous social value, considering its capacity to capture carbon and reintegrate nutrients into the soil. Another problem is that this type of innovation is usually produced over a longer period of time, and the commodification of science also produces, in an academic setting, a strong incentive to obtain narrow and short-term results (Radder 2010).

3.3.4 Easily Reproducible Innovations

Plant breeders who want to develop open-pollinated varieties, or other crops that are easily reseeded, will have difficulty finding funding for their research in an environment that assumes that innovators can recoup their research and development costs by developing marketable products (Kloppenburg 2005). Improvements of such varieties have value in themselves as they are crucial for maintaining biodiversity, add resilience to crop varieties, and can be used in buffer zones. Moreover, these improved varieties can be reseeded for subsequent harvests, allowing poorer

[2] Agreement on Trade-Related Aspects of Intellectual Property Rights (1996), art. 27(1).

farmers to benefit from scientific advancement and make a larger contribution to food security. Many farmer organizations also advocate for the reestablishment of the freedom to exchange seeds as part of their cultural understanding and customary practices, which would not only provide a valuable ecosystem service by expanding and maintaining agrobiodiversity, but would also allow the practice of the key social values of solidarity and reciprocity, as well as receiving recognition (Montenegro de Wit 2019). The prohibition of this practice has also entailed the loss of valuable forums to meet, exchange experiences and offer mutual help, particularly for rural women with few other opportunities to leave their homes (Shiva 2009). Here it is important to stress that, due to the poverty in which most smallholders live, innovations that can easily be reproduced are the ones most likely to be used and improve their situation (Gupta 2010).

3.3.5 *Validation of Old Data*

Proprietary science focuses on the production of new data, instead of facilitating the validation of old data (Leonelli 2007). Furthermore, the commercialization of innovation disincentivizes the sharing of data that may put sales in jeopardy. As a consequence, mistakes and fraud are difficult to detect once they are published in credible outlets. A classic example concerning food involves spinach: a calculation mistake had the effect that for decades parents forced their children to eat spinach under the false belief that this vegetable had huge iron content (Weigmann 2005).

3.4 Two Alternative Research Models Struggling for Survival

Perhaps the biggest negative effect of intellectual property on non-proprietary science is that it makes it difficult to keep parallel participatory research environments working. From an ethical perspective, it is particularly unfair that non-proprietary research models have to spend scarce resources to avoid the danger that innovation produced under their system becomes enclosed in proprietary science models, and thus ends up available only to those able and willing to purchase a license. A not always fail-safe way of protecting knowledge from entering the private domain is defensive publishing, which requires a sufficiently broad description of inventions to make clear to patent offices that there is "prior art", meaning that any similar invention filed for a patent should not count as novel, and should thus be ineligible (Sherman 2016). This clearly raises an issue of justice, as the cost created by intellectual property systems is also partly paid by those who reject these systems.

The existence of a proprietary science system that is able to generate its own resources puts non-proprietary science systems under pressure to justify their existence and *modus operandi*. The fact that some innovations can be sold profitably to recoup research and development costs has led to a conviction that innovation should be self-financing. This is particularly troublesome for open science organizations who do not even intend to develop products that can be sold in the market for profit.

To visualize the main problems of a scientific system dominated by intellectual property, we will now discuss in further detail two innovation systems that are currently struggling for their survival: traditional ecological knowledge systems and open science.

3.4.1 Traditional Ecological Knowledge Systems

The most widely-practiced way of doing science consists in the knowledge-gathering methods that fall under the broad category of traditional (ecological) knowledge systems. This category is very broad, and perhaps even attempting to include such a variety of knowledge-gathering methods under a single category is questionable. Here we find observation methods characterized by on-site observations, gathered by living or continuously working in the area where changes were implemented (Dutfield 2006). Much innovation is developed in incremental steps and is time-tested, often over several generations (Reyes-García et al. 2018; Robinson 2008). Large parts of traditional knowledge are not directly patentable, as they often do not meet the requirements of patentability. For instance, they may involve only a small inventive step, and therefore be counted as obvious by patent offices, or they may consist in public knowledge or have already been published, violating the novelty requirement. In many cases they may also be unsuitable for industrial application, as the invention may not be replicable using specific standards, since it may rely on adjustable ratios and relations. It is not uncommon for inventions directly based on traditional knowledge to meet none of these patent eligibility requirements. However, there have been many occasions when industry has relied on key genetic resources, identified in traditional knowledge, to develop patentable innovations (Soria-López and Fuentes-Páramo 2016; Blakeney 2019). Another problem is that, even if traditional knowledge can be patented, it might not be possible to identify the legitimate owner(s) of that knowledge, nor who should have control over the resource and its benefits (Moahi 2007).

Traditional knowledge systems have made a huge contribution to the advancement of science. There are strong voices in favour of maintaining such systems; for example, it is stated in the Declaration of Science and the Use of Scientific Knowledge (World Conference of Science 1999) "that traditional and local knowledge systems, as dynamic expressions of perceiving and understanding the world, can make, and historically have made, a valuable contribution to science and technology, and that there is a need to preserve, protect, research and promote this cultural heritage and empirical knowledge" (para. 26). Sometimes traditional knowledge is seen as a source of inspiration, and at other times, a repository of knowledge. One of the major struggles in this discourse has been to obtain recognition of the fact that traditional knowledge has both a static dimension that needs to be protected from disappearing, and a dynamic dimension which is continuously growing and needs to be promoted due to its knowledge-generating role (Robinson 2008).

A major benefit of improving and defending traditional knowledge systems is that they already include the largest number of participants. An improvement of communication and networking capacities among smallholders will therefore lead to a massive exchange, revision and implementation of traditional ecological knowledge in environmentally and socially dispersed and diverse environments. It would even allow the exchange of ideas and experiences between peoples who have never had the chance to collaborate before, and are currently overwhelmingly reliant on knowledge developed in very different socioeconomic and environmental contexts. Here techniques developed in one part of the world can be tested and adapted in other parts of the world. For instance, methods to protect crops from Caribbean hurricanes can be adapted to protect similar crops in East Asia from tropical cyclones.

The incorporation of traditional knowledge into the global scientific system is not without its problems. Independent observations have led to differences in taxonomy. Translating observations from one system to another is a scientific challenge, as distinctions get lost when separate groups of living organisms are treated as a single group, or when information about taxonomically different organisms is grouped together under a common name, leading to generalizations and risking ambiguity (Ludwig 2015). We may also find strong opposition to the instrumental treatment of knowledge when that knowledge is considered sacred by some communities.

Some have also accused scientists within the standard system of scientific imperialism and colonization efforts. This may have negative effects on efforts to incorporate traditional knowledge into the global scientific system, as there is a strong suspicion that this system has negative biases against such forms of knowledge. Yet it may also create an incentive to rediscover autochthonous knowledge systems and forgotten or underused local knowledge, and to demonstrate their potential and use them to fight prejudices, and inspire an element of pride among local communities. Another problem is that traditional ecological knowledge systems are not perceived by many as legitimate and valuable knowledge-acquisition approaches: something that might not only count as a form of discrimination when not backed by solid arguments, but may ultimately contribute to eroding their efficacy (Coolsaet 2015). Moreover, special attention needs to be paid to issues of biopiracy and the failure to acknowledge the contribution of traditional knowledge. These latter issues will be handled in more detail in Chap. 9, on restorative justice.

A major problem of traditional knowledge systems is that knowledge tends to be learned and transmitted only when it is of use to people, either for instrumental reasons or as heritage. If some generation does not find any use for traditional knowledge, it is likely that much of that knowledge will be lost. When indigenous communities can no longer find the species about which they have knowledge, they will be unlikely to pass this knowledge on to their descendants (Rozzi 2012). We are currently faced by a massive loss of such knowledge due to rapid urbanization, a lack of interest in rural work, and biases against traditional methods and in favour of industrialization (Robinson 2008). Documentation efforts involve a similar risk: traditional knowledge that is considered useful is likely to be prioritized in such efforts, while elements that are considered inessential may be discarded. Traditional knowledge and elements of that knowledge that fail standard scientific validation are also likely to be discarded (Agrawal 2002). As many such efforts are made by outsiders or with their collaboration, substantial losses and decontextualization will be experienced. Much knowledge is lost or underused when it does not serve the interests of those in power (Schiebinger 2004). Moreover, documentation can rarely conserve tacit knowledge, which may impede the replication of an invention in the future without the help of someone who makes practical use of that knowledge.

There is also the question of how traditional knowledge is to be managed. It could be protected under a specially-designed system of intellectual property (Liu and Gu 2011), or it could be considered part of the public domain, or it could be governed as a commons, following community rules (Reyes-García et al. 2018). Currently all three forms of traditional knowledge governance can be observed. Organizations like the Honeybee Network help grassroots innovators to seek intellectual property rights for the commercial use of their inventions, while at the same time encouraging knowledge exchange between peers (Gupta 2006; Shivarajan and Srinivasan 2013). In addition, traditional knowledge libraries have been established. Some of these efforts document knowledge and keep it restricted, so that traditional innovators may prove prior art to fight misappropriation. In other cases, these libraries are treated as commons and attempt to keep the knowledge in the public domain by using creative commons licenses to encourage diffusion and fair use, as the Spanish CONECT-e platform illustrates (Reyes-García et al. 2018).

3.4.2 Open Science Movements

To improve participation in scientific projects and their governance, a wide range of movements have emerged to advocate for open research and development environments. We find groups of people who value open science for both intrinsic and utilitarian reasons, that is, for the inherent value of doing science in a specific way and for the social benefits of an inclusive science system. This interest in openness, independently of whether it is pursued as a matter of rights, as a means to an end, or simply out of curiosity, has encouraged people and organizations from different backgrounds to work together. Here we find actors who have historically contributed to improving access to information, such as librarians and archivists. We also find actors from new professions, such as open source advocates, who have worked to improve access to source codes for software developers (Schoonmaker 2007). Major outrage at the rapid and enormous price spikes in academic and educational literature, together with the new opportunities that advances in information technologies offer, has brought together groups who work to facilitate wider access to scholarly literature, cultural heritage and educational resources in the open access movement (Atenas and Havemann 2014). A strong desire to conserve traditional seed exchange practices, and the need to conserve biodiversity and seed sovereignty, has triggered the development of open seed initiatives that aim to provide free access to materials in order to facilitate plant breeding and the cultivation of free varieties (Luby et al. 2016; Montenegro de Wit 2019), thus allowing a system based on reciprocity that encourages sharing (Kloppenburg 2010, 2014).

Access to scholarly literature has benefits for both scientists and the wider public. After financing much research through taxes, the public should have a right to see what has been done with their money (Anderson 2018). The wider public also needs access in order to monitor scientific advancement, to satisfy whatever curiosity they may have, to build an informed opinion about science and technology policies, to clarify their own questions and personal research needs, and in more general terms, to improve their own scientific literacy. Here it is particularly troublesome that only a very small percentage of the total number of works withheld by copyright laws for decades has any commercial value, but nevertheless uncommercialized works cannot be made public without fear of provoking infringement lawsuits (Boyle 2007).

One movement that is currently acquiring enormous momentum is open science (Leonelli 2017). This movement tries to revive and implement ideals of openness by embracing the potential that new information technologies have for bringing about this old goal, as well as the previously unimagined opportunities that an interconnected global scientific society offers, with English as a new *lingua franca* that has attracted an unprecedented number of speakers (cf. Van Parijs 2011). Intellectual property is seen as a major barrier to taking full advantage of this new chance to fully globalize scientific research networks.

Open science may offer opportunities to reuse findings, recover abandoned research lines, diversify publication venues, and encourage the use of high standards (Leonelli et al. 2015). It also invites researchers to share their data so that others can build on them.

An element of open science that is of special interest consists of efforts to produce community-owned scientific resources, such as databases, modelling plants and biomarkers. When such tools are held in a proprietary system they may allow unacceptable rent-seeking behaviour that slows down science. To fight this trend, open science advocates have started to develop openly accessible variants in order to establish those free versions as new research standards (Jefferson 2006).

Open innovation models have attracted interest in enhancing product development, in both the private and the public sector. Here these sectors have had to overcome two major predispositions. For one, they have had to leave aside traditional methods for marketing innovations and explore other ways to earn revenues. In addition, they have had to lose their tendency to be self-reliant and gain confidence in the quality, availability and capacity of the contributions of others (Chesbrough 2004). An open innovation system has the advantage of allowing people, irrespective of their status or origin, to share their expertise, opinions, questions and samples: something that contributes to social welfare and allows the establishment of collaborative communities with multiple benefits such as exercising solidarity and building social cohesion (Torrance and von Hippel 2015; Benkler and Nissenbaum 2006). This flexibility in allowing people to contribute to a scientific system facilitates epistemic diversity, as contributors will not have to be part of a school, institute or company to add their insights. It also allows people to contribute out of different motivations, without the hierarchies and rules of markets (Benkler 2017).

Open science has benefited from both legal and technological innovations. Advances in information and communication technologies have made the sharing of knowledge, know-how and information much cheaper and faster. It has also allowed people working in the same area to network, facilitating rapid communication and the sharing of knowledge and information. Innovators who work in regions with efficient postal services have also been able to collaborate by exchanging samples and prototypes. Legal innovations have also played a central role. One of the key accomplishments of open science movements has been to raise consciousness of the fact that property rights—including intellectual property rights—consist in a bundle of rights (cf. Honoré 1961). By separating intellectual property rights into a series of sub-rights, such as the right to attribution of authorship, the right to deny copying, the right to make derivative works, the right to object to commercialization, and the right to make usage conditional on sharing any changes on the same terms, open science advocates have given innovators the freedom to choose which rights they really want to conserve, and which rights they are not interested in (Deibel 2013; Ostrom and Hess 2007). To facilitate such choices and provide clarity about rights for both innovators, users and follow-up innovators, creative commons licenses were created (Deibel 2014). These types of license have contributed enormously to expanding the public domain. For example, if I take a photograph of a cassava farm in the Amazonas, I may want to share it freely with people for educational purposes, but at the same time would want something in return if it were used on a commercial basis. In the past, if I had been very hesitant to allow commercial use, I would have insisted on my copyright, which unfortunately would limit the freedom of others to use it for not-for-profit educational materials, collages and presentations. An added advantage of standardized licences, such as creative commons licenses, is that users only have to become familiar with their content and possibilities once, making subsequent uses much easier (Kierkegaard and Adrian 2010). Awareness of these different sub-rights allows people to withhold from the public domain only those uses that are dear to them. As a consequence, these works enter the public domain instantly and not long after their authors' death or the expiry of the relevant patents.

In order to stimulate participation, open collaboration networks would ideally have certain characteristics. They need to be modular in order to allow independent contributions, and also to allow incremental improvements (Schweik 2007). A well-known example is Wikipedia, which allows independent contributions by segmenting entries, and permits incremental improvements by facilitating instantaneous updates for all but protected entries (Kierkegaard and Adrian 2010). In agriculture, due to the enormous inequalities among research partners, it is important to not only pay attention to the modularity but also to the granularity of research objectives, by taking in consideration resource heterogeneity, socio-environmental circumstances, and the different disciplinary focusses and types of institutions (Basu et al. 2017). The huge advantage of research fields that allow independent contributions, or platforms designed to facilitate such contributions, is that they can be upscaled enormously, allowing contributions from around the globe. Research on genomes has benefited enormously from open science, as many institutions around the world have contributed to such projects. Since these models largely avoid property disputes, they facilitate the construction of valuable tools for future research that are usually too costly to be produced by single companies (Sá and Tamtik 2011).

Another tool that creative commons licenses offer is the share-alike clause. This clause has great potential to expand the public domain, by obliging users of resources licensed under this clause to contribute follow-up innovations under the same conditions under which the original was shared. By a pyramid effect, this widely increases the number of innovations that become freely available.

It is important to keep in mind that open science can work both as a type of commons and as an unrestricted public domain. In the first case, certain restrictions and community protocols can be established, for example to increase democratic decision-making and control over innovation processes (Kloppenburg 2010). Yet such processes can only be established if communities are built. Otherwise open science will merely consist in the aggregation of contributors working in anarchy, leaving the direction of research to be determined by individual curiosity, without any commonly established goals.

In agriculture, such models have great potential, as farmers and scientists can contribute their insights and worries from a wide variety of environmental, social, methodological and scientific backgrounds. Open innovation systems may help to identify, at very early stages, potential and secondary uses, possible threats, errors, and social and environmental needs. They also improve the unbiased monitoring of the performance of an invention, as test results from diverse social and environmental settings can be made available. Open data can also reveal information on the areas of the world in most need of agricultural innovation, by providing information on crop, yields, farm size and technology uptake (Ricciardi et al. 2018). Beside its scientific value, encouraging participation is something people value in itself. Openness is also particularly important for agrobiodiversity (Prathapan and Rajan 2011). Open seed banks facilitate in situ conservation and allow breeders to draw from a more diverse gene pool. In contrast to other types of natural resource, sharing and using does not erode genetic resources. It is rather the underuse of these resources which threatens their conservation, further improvement and availability (Halewood 2013).

Yet open innovation models are not without their problems. A problem of special concern is the distribution of freely available innovations (Louwaars 2007). If nobody earns money from distributing a resource, this may have the result that the inventions end up being underused even when they are freely available, since product promotion and distribution costs money. Similarly, many products concerning biological materials need to pass certain biosafety or certification tests before they can be widely distributed. Such tests cost money and pose a serious hurdle for the public to benefit from non-proprietary scientific advancement. Moreover, open science needs to provide some securities, especially to avoid scooping by well-funded research

Fig. 3.1 The different degrees of controlling a good through property rights

institutes and the over-criticism of still premature research lines (Leonelli et al. 2015).

Another problem is that innovators who contribute to open systems rarely get a fair reward for their contribution. The public benefit of openness regularly comes at the cost of failing to reward people for their work. These systems may fail to remunerate innovators in any form, be it through public subventions, prizes or donations. However, there are exceptions when it comes to payments within open innovation models. In the software industry, open source software has drawn a distinction between open and free software—corresponding to the difference in Spanish between "*libre*" and "*gratis*"—to highlight the difference between being free of boundaries or limitations for further innovation and being available without cost. In practice, this has led private companies to market open source software with specially tailored add-ons and support packages, so as to create an end product that people will be willing to pay for. Knowledge-intensive agriculture could benefit from similar strategies.

To wrap up, it should be noted that there is no rigid dichotomy between proprietary innovation and open innovation (see Fig. 3.1). In practice, business models and research communities set their innovation management policies within a spectrum that ranges from very closed to very open systems (Tietze et al. 2017), ranging from exclusive licenses to creative commons licenses that only retain attribution of authorship.

3.5 General Problem: The Diffusion and Adoption of Technologies

The diffusion and proper adoption of technologies is a major challenge for improving agriculture (Klerkx et al. 2012; Röling et al. 1976). Innovations outside the proprietary system need substantial support from governments and organizations to be adopted by farmers. This requires specialists in agricultural extension and the help of farmers who are willing to cooperate and assist in the adoption of the new methods. Here horizontal knowledge-exchange practices, such as the *campesino-a-campesino* movement, have shown great potential for diffusing knowledge and fostering networks of solidarity (Rosset et al. 2011; Holt-Giménez 2006). The networks needed to diffuse such improvements can also be used to explore different cooperation opportunities, such as developing alternative sales networks and organizing political activism.

3.6 Aims of Justice

As mentioned before, science and technology development are not morally neutral. Non-proprietary science has advantages and disadvantages for social justice that need to be discussed. Here is a brief overview, organized according to the six dimensions of social justice, which will help us to understand some of the key challenges that agricultural innovation faces. These challenges will be discussed in more detail in the following chapters.

The access problem and justice in exchange. Documented innovation which is not covered by intellectual property rights is free to be used. This has the significant advantage that it can be used by all who are interested. Unfortunately, when nobody profits from selling such innovations, they are likely to be insufficiently diffused. Some companies have started to provide additional services, for example tailoring openly-accessible technologies to local needs or offering training, in order to gain financially from diffusing such technologies. This increases access, at least for those with a few resources to spare. Unfortunately, the majority of innovators who contribute to open science tend not to be rewarded for their efforts, even when society obtains major benefits from their work.

The availability problem and distributive justice. A central criticism of non-proprietary science systems is that they do not sufficiently incentivize investments in innovation. This argument ignores alternative funding mechanisms, such as prizes, but nevertheless highlights an important problem: non-proprietary science is dependent on outside funding or people's spare time. The fact that innovators do not rely on market incentives allows them to concentrate on scientifically underserved groups.

Facilitating participation and contributive justice. The greatest advantage of non-proprietary science is that it allows wide-scale participation. This has, as mentioned before, benefits for participants and for those who benefit from the fruits of scientific advancement. Yet here we should not forget that current levels of deprivation in nutrition and scientific literacy are, in terms of numbers, far greater impediments than intellectual property rights when it comes to hindering user innovation.

Encouraging sustainability and intergenerational justice. There is no strong causal link between open science and environmental sustainability. Additional efforts have to be made so that environmentally friendly technologies are developed and distributed. The free availability of inventions does not automatically lead to their adequate diffusion. In particular, inventions that require training and tacit knowledge for their operation require substantial efforts for their uptake.

Aligning good scientific practice and procedural justice. Non-proprietary science can emerge from public institutions and thereby be accountable to taxpayers, yet innovations can also emerge from self-funded individuals and communities, over whose actions society has little say, raising issues of accountability similar to those seen in corporate science.

Arbitration of disputes and restorative justice. By appealing to openness and common ownership, some of the key problems of science are prevented. Yet other problems, such as failing to obtain informed consent, or preventing other forms of scientific malpractice, remain unresolved. This is alarming, since scientists who are convinced that they are promoting wider social interests may ignore issues of consent by following utilitarian considerations.

3.7 Concluding Remarks

From this rough overview, we have obtained a better idea of which kinds of innovation system and which types of innovation are endangered by overly relying on intellectual property rights. Taking into account the advantages and disadvantages of these alternative systems, we can better assess the demands of social justice specified in the following chapters.

References

Agrawal, Arun. 2002. Indigenous knowledge and the politics of classification. *International Social Science Journal* 54 (173): 287–297.

Altieri, Miguel A. 2002. Agroecological principles for sustainable agriculture. In *Agroecological Innovations: Increasing Food Production with Participatory Development*, ed. Norman Uphoff, 40–46. London & Sterling, VA: Earthscan.

Anderson, Molly D. 2018. Scientific knowledge of food and agriculture in public institutions: movement from public to private goods. In *Routledge Handbook of Food as a Commons*, ed. José Luis Vivero-Pol, Tomaso Ferrando, Olivier De Schutter and Ugo Mattei, 185–202. Oxon & New York: Routledge.

Atenas, Javiera, and Leo Havemann. 2014. Questions of quality in repositories of open educational resources: A literature review. *Research in Learning Technology* 22: 20889.

Basu, Soutrik, Joost Jongerden, and Guido Ruivenkamp. 2017. The emergence of a hybrid mode of knowledge production in the Generation Challenge Programme Rice Research Network (GCP-RRN) in India: Exploring the concept of Commons-Based Peer Production (CBPP). *Geoforum* 84: 107–116.

Benkler, Yochai. 2017. Peer production, the commons, and the future of the firm. *Strategic Organization* 15 (2). 264–274.

Benkler, Yochai, and Helen Nissenbaum. 2006. Commons-based peer production and virtue. *The Journal of Political Philosophy* 14 (4): 394–419.

Besson, Samantha. 2015. Science without borders and the boundaries of human rights. *European Journal of Human Rights* 4: 462–485.

Blakeney, Michael. 2019. Remedying the Misappropriation of Genetic Resources. In *Intellectual Property Issues in Microbiology*, ed. Harikesh Bahadur Singh, Chetan Keswani and Surya Pratap Singh, 147–161. Singapore: Springer.

Boyle, James. 2007. Mertonianism Unbound?: Imagining free, decentralized access to most cultural and scientific material. In *Understanding Knowledge as a Commons*, ed. Charlotte Hess and Elinor Ostrom, 123–143. Cambridge, MA & London: MIT Press.

Boyle, James. 2008. *The Public Domain. Enclosing the Commons of the Mind.* New Haven & London: Yale University Press. http://thepublicdomain.org.

Chesbrough, Henry. 2004. Managing open innovation. *Research-Technology Management* 47 (1): 23–26.

Coolsaet, Brendan. 2015. Transformative participation in agrobiodiversity governance: Making the case for an environmental justice approach. *Journal of Agricultural and Environmental Ethics* 28 (6): 1089–1104.

Deibel, Eric. 2013. Open variety rights: Rethinking the commodification of plants. *Journal of Agrarian Change* 13 (2): 282–309.

Deibel, Eric. 2014. Open genetic code: On open source in the life sciences. *Life Sciences, Society and Policy* 10: 2.

di Norcia, Vincent. 2005. Intellectual property and the commercialization of research and development. *Science and Engineering Ethics* 11 (2): 203–219.

Dutfield, Graham. 2006. Protecting traditional knowledge: Pathways to the future. In *ICTSD Programme on IPRs and Sustainable Development.* Geneva: International Centre for Trade and Sustainable Development.

Eisenberg, Rebecca S., and Richard R. Nelson. 2002. Public vs. proprietary science: A fruitful tension? *Daedalus* 131 (2): 89–101.

Fernández Pinto, Manuela. 2015. Commercialization and the limits of well-ordered science. *Perspectives on Science* 23 (2): 173–191.

Geertsema, Willemien, Walter A.H. Rossing, Douglas A. Landis, Felix J.J.A. Bianchi, Paul C.J. Rijn, Joop H.J. Schaminée, Teja Tscharntke, and Wopke Werf. 2016. Actionable knowledge for ecological intensification of agriculture. *Frontiers in Ecology and the Environment* 14 (4): 209–216.

Gilson, Erinn Cunniff. 2015. Vulnerability, relationality, and dependency: Feminist conceptual resources for food justice. *IJFAB: International Journal of Feminist Approaches to Bioethics* 8 (2): 10–46.

Gliessman, Stephen R. 2007. *Agroecology: The Ecology of Sustainable Food Systems.* Boca Raton: CRC Press.

Goldenberg, Maya J. 2016. Public misunderstanding of science? Reframing the problem of vaccine hesitancy. *Perspectives on Science* 24 (5): 552–581.

Grey, Sam, and Raj Patel. 2015. Food sovereignty as decolonization: Some contributions from Indigenous movements to food system and development politics. *Agriculture and Human Values* 32 (3): 431–444.

Gupta, Anil K. 2006. From sink to source: The Honey Bee Network documents indigenous knowledge and innovations in India. *Innovations* (summer): 49–66.

Gupta, Anil K. 2010. Grassroots green innovations for inclusive, sustainable development. In *The Innovation for Development Report 2009–2010, Strengthening Innovation for the Prosperity of the Nations*, ed. Augusto, Lopez-Claros, 137–146. Houndmills & New York: Palgrave Macmillan.

Halewood, Michael. 2013. What kind of goods are plant genetic resources for food and agriculture? Towards the identification and development of a new global commons. *International Journal of the Commons* 7 (2): 278–312.

Hess, Charlotte, and Elinor Ostrom. 2007. *Understanding Knowledge as a Commons.* Cambridge, MA & London: MIT Press.

Holt-Giménez, Eric. 2006. *Campesino a campesino: Voices from Latin America's farmer to farmer movement for sustainable agriculture.* Oakland: Food First Books.

Honoré, Anthony M. 1961. Ownership. In *Oxford Essays in Jurisprudence*, ed. A. Guest, 107–147. London: Oxford University Press.

Intemann, Kristen, and Inmaculada de Melo-Martín. 2014. Addressing problems in profit-driven research: How can feminist conceptions of objectivity help? *European Journal for Philosophy of Science* 4 (2): 135–151.

Jefferson, Richard A. 2006. Science as social enterprise: The CAMBIA BiOS initiative. *Innovations: Technology, Governance, Globalization* 1 (4): 13–44.

Kelbessa, Workineh. 2015. African environmental ethics, indigenous knowledge, and environmental challenges. *Environmental Ethics* 37 (4): 387–410.

Kierkegaard, Patrick, and Angela Adrian. 2010. Wikitopia: Balancing intellectual property rights within open source research databases. *Computer Law & Security Review* 26 (5): 502–519.

Klerkx, Laurens, Barbara Van Mierlo, and Cees Leeuwis. 2012. Evolution of systems approaches to agricultural innovation: Concepts, analysis and interventions. In *Farming Systems Research into the 21st Century: The New Dynamic*, ed. Ika Darnhofer, David Gibbon, and Benoît Dedieu, 457–483. Dordrecht: Springer.

Kloppenburg, Jack. 2005. *First the Seed: The Political Economy of Plant Biotechnology*, 2nd ed. Madison: University of Wisconsin Press.

Kloppenburg, Jack. 2010. Impeding dispossession, enabling repossession: Biological open source and the recovery of seed sovereignity. *Journal of Agrarian Change* 10 (3): 367–388.

Kloppenburg, Jack. 2014. Re-purposing the master's tools: The open source seed initiative and the struggle for seed sovereignty. *Journal of Peasant Studies* 41 (6): 1225–1246. https://doi.org/10.1080/03066150.2013.875897.

Koepsell, David. 2010. Back to basics: How technology and the open source movement can save science. *Social Epistemology* 24 (3): 181–190.

Leonelli, Sabina. 2007. Growing weed, producing knowledge an epistemic History of *Arabidopsis thaliana*. *History and Philosophy of the Life Sciences* 29 (2): 193–223.

Leonelli, Sabina. 2017. *MLE on Open Science: Altmetrics and Rewards—Incentives and Rewards to engage in Open Science Activities*. Brussels: European Commission.

Leonelli, Sabina, Daniel Spichtinger, and Barbara Prainsack. 2015. Sticks and carrots: Encouraging open science at its source. *Geo: Geography and Environment* 2 (1): 12–16.

Liu, Changhua, and Gu Man. 2011. Protecting traditional knowledge of Chinese medicine: Concepts and proposals. *Frontiers of medicine* 5 (2): 212–218.

Louwaars, Niels. 2007. *Seeds of confusion: The impact of policies on seed systems*: Ph.D. thesis, Wageningen Universiteit.

Luby, Claire H., Jack R. Kloppenburg, and Irwin L. Goldman. 2016. Open source plant breeding and the open source seed initiative. *Plant Breeding Reviews* 40: 271–298.

Lucier, Paul. 2019. Can marketplace science be trusted? *Nature* 574: 481–485.

Ludwig, David. 2015. Indigenous and scientific kinds. *British Journal for the Philosophy of Science* 68 (1): 187–212.

Macfarlane, Bruce, and Ming Cheng. 2008. Communism, universalism and disinterestedness: Re-examining contemporary support among academics for Merton's scientific norms. *Journal of Academic Ethics* 6 (1): 67–78. https://doi.org/10.1007/s10805-008-9055-y.

May, Christopher. 2006. Patents, universities and the provision of social goods in the information society. *Ethical Perspectives* 13 (2): 289–304.

Merton, Robert K. 1942. Science and technology in a democratic order. *Journal of Legal and Political Sociology* 1 (1–2): 115–126.

Moahi, Kgomotso H. 2007. Globalization, knowledge economy and the implications for indigenous knowledge. *International Review of Information Ethics* 7: 55–62.

Montenegro de Wit, Maywa. 2019. Beating the bounds: How does 'open source' become a seed commons? *The Journal of Peasant Studies* 46 (1): 44–79.

Ostrom, Elinor, and Charlotte Hess. 2007. A framework for analyzing the knowledge commons. In *Understanding Knowledge as a Commons*, ed. Charlotte Hess and Elinor Ostrom, 41–82. Cambridge, MA & London: MIT Press.

Peres, Sara. 2016. Saving the gene pool for the future: Seed banks as archives. *Studies in History and Philosophy of Science Part C: Studies in History and Philosophy of Biological and Biomedical Sciences* 55: 96–104. https://doi.org/10.1016/j.shpsc.2015.09.002.

Piesse, Jenifer, and C. Thirtle. 2010. Agricultural R&D, technology and productivity. *Philosophical Transactions of the Royal Society B: Biological Sciences* 365 (1554): 3035–3047.

Plumwood, Val. 2006. The concept of a cultural landscape: Nature, culture and agency in the land. *Ethics and the Environment* 11 (2): 115–150.

Prathapan, K.Divakaran, and Priyadarsanan Dharma Rajan. 2011. Biological diversity: A common heritage. *Economic & Political Weekly* 46 (14): 15–17.

Preston, Christopher J., and Fern Wickson. 2016. Broadening the lens for the governance of emerging technologies: Care ethics and agricultural biotechnology. *Technology in Society* 45, 48–57.

Radder, Hans. 2010. The commodification of academic research. In *Science and the Modern University*, ed. Hans Radder, 1–23. Pittsburg: University of Pittsburg Press.

Raymond, Eric. 1999. The cathedral and the bazaar. *Knowledge, Technology & Policy* 12 (3): 23–49.

Reyes-García, Victoria, Petra Benyei, and Laura Calvet-Mir. 2018. Traditional agricultural knowledge as a commons. In *Routledge Handbook of Food as a Commons*, ed. José Luis Vivero-Pol, Tomasso Ferrando, Olivier De Schutter and Ugo Mattei, 173–184. Oxon & New York: Routledge.

Ricciardi, Vincent, Navin Ramankutty, Zia Mehrabi, Larissa Jarvis, and Brenton Chookolingo. 2018. How much of the world's food do smallholders produce? *Global Food Security* 17: 64–72.

Robinson, Daniel. 2008. Beyond 'protection': Promoting traditional knowledge systems in Thailand. In *Patenting Lives: Life Patents, Culture and Development*, ed. Johanna Gibson, 121–138. Aldershot: Ashgate Publishing.

Röling, Niels G., Joseph Ascroft, and Fred Wa Chege. 1976. The diffusion of innovations and the issue of equity in rural development. *Communication Research* 3 (2): 155–170.

Rosset, Peter Michael, Braulio Machin Sosa, Adilén María Roque Jaime, and Dana Rocío Ávila Lozano. 2011. The Campesino-to-Campesino agroecology movement of ANAP in Cuba: Social process methodology in the construction of sustainable peasant agriculture and food sovereignty. *The Journal of Peasant Studies* 38 (1): 161–191.

Rozzi, Ricardo. 2012. Biocultural ethics: Recovering the vital links between the inhabitants, their habits, and habitats. *Environmental Ethics* 34 (1): 27–50.

Sá, Creso, and Merli Tamtik. 2011. Structural genomics and the organisation of open science. *Genomics, Society and Policy* 7 (1): 20–34.

Schiebinger, Londa. 2004. Feminist history of colonial science. *Hypatia* 19 (1): 233–254.

Schoonmaker, Sara. 2007. Globalization from below: Free software and alternatives to neoliberalism. *Development and Change* 38 (6): 999–1020.

Schweik, Charles M. 2007. Free/open-source software as a framework for establishing commons in science. In *Understanding Knowledge as a Commons*, ed. Charlotte Hess and Elinor Ostrom. Cambridge, MA & London: MIT Press.

Shackelford, Scott J. 2008. The tragedy of the common heritage of mankind. *Stanford Environmental Law Journal* 27: 101–157.

Shaver, Lea. 2010. The right to science and culture. *Wisconsin Law Review* 121: 121–184.

Sherman, Brad. 2016. Reconceptualizing intellectual property to promote food security. In *The Intellectual Property and Food Project*, ed. Charles Lawson and Jay Sanderson, 39–54. London: Routledge.

Shiva, Vandana. 2009. Women and the gendered politics of food. *Philosophical Topics* 37 (2): 17–32.

Shivarajan, Sridevi, and Aravind Srinivasan. 2013. The poor as suppliers of intellectual property: A social network approach to sustainable poverty alleviation. *Business Ethics Quarterly* 23 (3): 381–406.

Soria-López, Manuel, and Israel Fuentes-Páramo. 2016. The identification of biopiracy in patents. *World Patent Information* 47: 67–74.

Steinvorth, Ulrich. 2009. The right to work and the right to develop one's capabilities. *Analyse & Kritik* 1: 101–113.

Stephan, Paula E. 2012. *How Economics Shapes Science*. Cambridge, MA: Harvard University Press.

Sterckx, Sigrid. 2011. Patenting and licensing of university research: Promoting innovation or undermining academic values? *Science and Engineering Ethics* 17 (1): 45–64.

Stiglitz, Joseph E. 1999. Knowledge as a global public good. In *Global Public Goods: International Cooperation in the 21st Century*, ed. Inge Kaul, Isabelle Grunberg and Marc A. Stern, 308–325. New York & Oxford: Oxford University Press.

Šūmane, Sandra, Ilona Kunda, Karlheinz Knickel, Agnes Strauss, Talis Tisenkopfs, Ignacio des Ios Rios, Maria Rivera, Tzruya Chebach, and Amit Ashkenazy. 2018. Local and farmers' knowledge matters! How integrating informal and formal knowledge enhances sustainable and resilient agriculture. *Journal of Rural Studies* 59: 232–241.

Taubman, Antony. 2005. Saving the village: Conserving jurisprudential diversity in the international protection of traditional knowledge. In *International Public Goods and Transfer of Technology Under a Globalized Intellectual Property Regime*, ed. Keith E. Maskus and Jerome H. Reichman, 521–564. Cambridge & New York: Cambridge University Press.

Tietze, Frank, Elisabeth Eppinger, Jan Sternkopf, and Pratheeba Vimalnath. 2017. IP strategies for sustainability, in *2017 IEEE Technology & Engineering Management Conference (TEMSCON)*. Santa Clara: IEEE. https://doi.org/10.1109/TEMSCON.2017.7998373.

Timmermann, Cristian. 2014. Sharing in or benefiting from scientific advancement? *Science and Engineering Ethics* 20 (1): 111–133.

Timmermann, Cristian, and Georges F. Félix. 2015. Agroecology as a vehicle for contributive justice. *Agriculture and Human Values* 32 (3): 523–538.

Timmermann, Cristian, and Zoë Robaey. 2016. Agrobiodiversity under different property regimes. *Journal of Agricultural and Environmental Ethics* 29 (2): 285–303. https://doi.org/10.1007/s10806-016-9602-2.

Timmermann, Cristian, and Henk van den Belt. 2012. Global justice considerations for a proposed "Climate Impact Fund". *Public Reason* 4 (1–2): 182–196.

Tittonell, Pablo. 2013. *Farming Systems Ecology: Towards Ecological Intensification of World Agriculture*. Wageningen: Wageningen Universiteit.

Torrance, Andrew W., and Eric von Hippel. 2015. The right to innovate. *Michigan State Law Review* 2015 (2): 793–829.

Trerise, Jonathan. 2016. The influence of patents on science. *Politics, Philosophy & Economics* 15 (4): 424–450.

van den Belt, Henk. 2010. Robert Merton, intellectual property, and open science. In *The Commodification of Academic Research: Science and the Modern University*, ed. Hans Radder, 187–230. Pittsburgh: University of Pittsburgh Press.

Van Parijs, Philippe. 2011. *Linguistic Justice for Europe and for the World*. Oxford; New York: Oxford University Press.

Vanloqueren, Gaëtan, and Philippe V. Baret. 2009. How agricultural research systems shape a technological regime that develops genetic engineering but locks out agroecological innovations. *Research Policy* 38 (6): 971–983.

Weigmann, Katrin. 2005. The consequence of errors. *EMBO Reports* 6 (4): 306–309. https://doi.org/10.1038/sj.embor.7400389.

Wickson, Fern. 2016. Do we care about synbiodiversity? questions arising from an investigation into whether there are GM crops in the Svalbard global seed vault. *Journal of Agricultural and Environmental Ethics* 29 (5): 787–811.

Wolfrum, Rüdiger. 1983. The principle of the common heritage of mankind. *Heidelberg Journal of International Law* 43: 312–337.

World Conference of Science. 1999. Declaration of Science and the Use of Scientific Knowledge. Budapest: UNESCO. http://www.unesco.org/science/wcs/eng/declaration_e.htm.

Chapter 4
Agricultural Innovation and Justice in Exchange: Beyond Fair Rewards and Adequate Access

Abstract Striking a balance between fair rewards and the provision of adequate access is one of the key social challenges of innovation systems. To illustrate the complexities involved in the commercialization of agricultural research, I examine in this chapter the different demands that justice in exchange sets for innovation systems. I start by discussing the foundational question of whether people are entitled to commercialize the inventions they control. After that, I introduce the central demands of justice in exchange: (i) informed consent, (ii) the prohibition of deceitful practices, (iii) proportionality of rewards, (iv) avoiding harmful effects, and (v) recognition.

Keywords Justice in transaction · Trade · Commercialization · Rent-seeking · Private and public interests

4.1 Introduction

Traditionally, the concept of justice in exchange has been concerned with how to make the transaction between two goods fairer or more equitable. It builds on a distinction drawn by Aristotle in his Nicomachean Ethics, between *distributive justice*, concerning what people can legitimately claim from shared resources, and *commutative justice*, concerning the circumstances under which transactions can be assessed as fair. Aristotle also distinguishes these latter two forms of justice from matters that need to be handled ex post as issues of *corrective justice*. In the more recent literature, this concept of justice tends to be called "justice in transactions", "transactional justice", or as I will say here, "justice in exchange" (Schroeder and Pisupati 2010; Koller 2013).

In order to consider a transaction as just, justice in exchange sets a wide set of demands. As a first issue, justice in exchange is concerned with the general questions of how people become entitled to legitimately exchange certain goods, and of whether certain goods should not be exchanged at all. Second, justice in exchange deals with the conditions that have to be met to ensure that transactions are carried out fairly. Here, we can distinguish between a narrow and broad understanding of this concept. In a narrow sense, justice in exchange seeks informed consent from the transacting

C. Timmermann, *Social Justice and Agricultural Innovation*, The International Library of Environmental, Agricultural and Food Ethics 31,
https://doi.org/10.1007/978-3-030-56193-2_4

parties, an abstinence from requesting disproportional remunerations, that exchanges do not lead to avoidable harm, and condemns deceit and the use of power to impose abusive exchange conditions unilaterally (Walton 2014; Koller 2013). A broader understanding of this concept recognizes that exchanges between two parties may have consequences that extend far beyond those directly involved in the transaction. In this sense, just transactions should avoid as far as possible indirectly harming others.

A well-developed concept of justice in exchange, in either its narrow or its wider interpretation, can provide a normative framework for arguing for increased access to the objects of innovation, and for condemning commercial behaviour that has harmful social and environmental consequences. This is particularly important, since the great need for agricultural innovation makes the commercialization of this type of innovation particularly problematic, as it may increase the difficulty of securing sufficiently large harvests to meet food needs and of reducing our environmental footprint. Relying on markets for technology diffusion is even more worrisome, if we consider the extreme inequality that can be observed both between and within different transacting parties, in terms of purchasing power, genetic resources, scientific infrastructure and harvest yields (Pogge 2008; Tittonell et al. 2016).

When we talk about a fair exchange, we should also consider people's individual perception of exchanges, which they may directly or indirectly experience—even if based on insufficiently informed or spontaneous decisions. As mentioned before, there is evidence that people will reduce their collaboration or even stop cooperating altogether when they perceive a system as unjust (Ooms 2010). This is also the case when it would be more economically rational to continue cooperating: the urge to punish group members who deviate from social norms or act wrongfully is often stronger than the desire for further rewards (Dodds 2005). Unfortunately, a perception of injustice can also be based on misconceptions, false beliefs or manipulation, and thereby affect people's willingness to cooperate. It is important to notice that a lack of participation in the conservation and creation of public goods—such as scientific knowledge and agrobiodiversity—is something that affects us all. Educational campaigns are needed to encourage participation in common enterprises, and rectificatory measures are needed to handle injustices (see Chap. 9).

Also relevant for justice is the case when a group that has suffered historical discrimination tends to consequently find a slightly better exchange as fair, even though it is still objectively unfair, just because it is the best arrangement that the disadvantaged group hopes to achieve in light of their historical experience. The phenomenon where disadvantaged groups are satisfied with slight improvements, which objectively still count as unfair, has been described as "adaptive preferences". This consists in an acquired coping mechanism whereby one limits one's preferences to what one believes one can realistically achieve under one's socio-economic circumstances or background conditions (Dieterlen 2003; Sen 2009). People who have lowered their expectations are much more vulnerable to accepting exploitative agreements, and will be less likely to identify their unfair nature.

As far as global justice is concerned, justice in exchange mostly deals with rules and procedures that have to be followed in order to increase the chances of making a commercial transaction fair for all affected parties. Some of these rules can be justified on the basis of general reciprocity. The demands of justice in exchange do not allow treating compatriots differently from distant people when it comes to condemning deception, the use of force or exploitation. This does, however, leave significant room for choosing trading partners, thus raising tensions between cosmopolitans and nationalists, and between advocates of economic development and advocates of environmental sustainability. A widely supported movement in favour of local producers is the food sovereignty movement (Nyéléni Forum for Food Sovereignty 2007).

Taking into account both perceived and actual injustices, in this chapter, I discuss the main demands of justice in exchange regarding the commercialization and trade involved in agricultural innovation. First, before being able to talk about fair exchanges, I need to examine whether people are entitled to exchange the objects they claim to own, or more generally, whether they have full property rights over such goods. Second, I discuss issues of informed consent concerning the purchase and sale of objects of innovation and genetic resources. Third, I define deceit as it applies to our field of analysis. Fourth, I examine the difficulties involved in identifying what amounts to an adequate remuneration. Fifth, I discuss the problem of avoiding harm, both directly and indirectly, in a world of limited resources where people compete for basic goods. Sixth, I mention some issues regarding the importance of recognition in trade. Finally, I make some observations on whether we should have a market for agricultural innovation at all.

4.2 On Being Entitled to Exchange

To start, the most general question we need to ask ourselves is who is entitled to negotiate over the goods in question. Who—if anyone—has full property rights over the objects of innovation? Is this person or institution allowed to transfer all of these underlying property rights? Such questions are particularly problematic for agricultural innovation, especially in plant breeding, as this relies on the work of countless generations dispersed through different geographic locations (Kloppenburg 2005; Mazoyer and Roudart 2006). This makes questions of who is the legitimate owner difficult and in some cases impossible to settle from a moral perspective.

First, there is a problem with the number of people claiming property rights. Individual ownership models do not sufficiently acknowledge historical and parallel contributions, yet they have the advantage that it is usually easier to bargain with a single owner. Communal ownership acknowledges a wider set of contributors, but may make negotiations more problematic, occasionally to the point of stagnation, as the different owners may have conflicting interests and expectations. Given the extensiveness of the genetic pool and the related knowledge that innovators need to draw on to develop new products, some authors regard the use of intellectual property rights as generally problematic, due to the large numbers of owners with conflicting interests who end up blocking each other, creating a problem of "hyperownership"

(Safrin 2004) or a "tragedy of the anti-commons" (Heller 2008). This is particularly problematic when there is no clarity over property boundaries, concerning who owns what and where one person's property begins and another's ends, as is often the case with patents for genetic resources.

The epistemic difficulty involved in singling out varieties and the scope of an improvement is also a major problem. If a single community negotiates access and benefit-sharing agreements with a multinational corporation in relation to a particular genetic resource, other communities who have contributed towards its improvement by being historically involved in its selection, breeding and stewardship, may end up empty-handed. When many communities have access to this genetic resource, some communities may act individually to maximize their opportunities to gain additional income by rushing to make the best offer, in the fear that another community will make an offer more convenient to the company's interests first. This risk is high when the seed sector is highly consolidated (ETC Group 2008; Wield et al. 2010) and biodiversity-rich communities do not act as a single interest group, losing bargaining power as a consequence of this fragmentation. But here again we face the earlier question: is a single community entitled to make commercial agreements over a good when other communities were also involved in its development?

When a single state can legitimately claim ownership over a genetic resource that can only be found in their territory, many of the problems of bargaining and collective agreement disappear. It is important to note here that few states display exemplary conduct towards the indigenous communities and smallholders within their jurisprudence, and it is exactly these groups who have played a key role in the conservation and further development of agrobiodiversity and traditional knowledge. Moreover, while some states have sufficient bargaining power to oppose multinational corporations, many megabiodiverse states are the weaker parties in negotiations with global corporations, in terms of both financial resources and professional expertise.

Secondly, there is the question of legitimacy. Did the negotiating parties acquire the object legitimately? Are historical injustices involved? Ideally, only legitimate owners are entitled to trade their resources. There are many cases of questionable acquisition of traditional knowledge and genetic resources in agricultural innovation. One problem, for instance, is that heads of state without democratic legitimacy are usually perceived by the international community as entitled to negotiate over the natural resources of the territory they exert sovereignty over: facilitating corruption and the exploitation of resources, against local citizens' interests (Pogge 2008; Wenar 2015). In addition, there is the problem of historical injustices. In general terms, issues of original acquisition are extremely difficult and often impossible to settle when addressed after longer periods of time (Waldron 2003; Patton 2005). Descendants of people who illegitimately acquired or appropriated a resource from which innovations are made (germplasm or traditional knowledge) have mixed these resources with a substantial amount of labour of their own. New social and environmental circumstances may also justify a new type of claim on these resources, based on necessity (Waldron 1992). We may claim access to an innovation because we need it in order to secure basic needs that we could not otherwise secure. While these factors may justify or at least morally excuse a right to use based on needs, they do not justify a right to seek disproportionate rents from others. The latter improvements also deserve compensation, particularly if we as a community benefit from them.

Third, there is the long-standing public outcry over the commercialization of research done with public money. One line of complaint is that public institutions should promote public science and work towards developing key research tools and foundational science by not engaging in commercial exploitations, or at least offering licences on generous terms (May 2006). Moreover, taxpayers are unwilling to pay twice for research, as happens when they pay the costs of research by financing public institutions yet also end up paying to access the fruits of such research, with substantial surcharges on top of cost prices (Standing 2016). Depending on one's political views, one might argue that public institutions should focus on directly serving the general citizenship and not concentrate on supporting industry, as the latter may only indirectly benefit the common good at the risk of increasing inequalities and creating environmental and health hazards.

Fourth, there may be general public interest clauses prohibiting the transfer of property rights. Here there are biosafety concerns about protecting public health and the environment, and also public safety concerns, when the innovation or genetic germplasm involved may be used to develop a weapon, or a threat to public health or other social interests.

Fifth, similar public interest clauses defend keeping a resource in the public domain, or in some cases within a community or state. Such interests may vary over time. Some generations, for example, may have forfeited significant financial rewards by not privatizing a resource, while others may have invested a large amount of public resources in stewarding these resources as state property, thus retaining democratic control. It would be unfair that in times of weak institutions or widespread opportunism among politicians, the public loses control over these resources and future generations end up paying the price of such enclosure, with little or no benefit to themselves. The nowadays common practice where elected political officials can freely privatize collective resources, with enormous effects that reach far beyond their legislation period, are from a moral and democratic perspective particularly questionable.

Finally, we should note that the readiness of patent offices to recognize as the inventor the first applicant who comes with a well-documented invention to a patent office, and award her exclusive rights, is not solely backed by notions of desert. The creation and maintenance of commercial incentives plays a key role in defending this practice. Here it is claimed that innovators and their financers need financial security for their investments, and that over-questioning ownership may disincentivize people from investing in research. Ensuring higher investments in research may thus come at the price of failing to reward some key innovators.

Due to the different factors affecting the ethical legitimacy of whoever becomes entitled to exchange a given resource, we need to assess exchanges on a case-by-case basis, keeping in mind that we may have to allow certain injustices for the sake of maintaining scientific advancement.

4.3 Informed Consent for Transactions

One of the factors that determine whether a transaction is fair is whether the transacting parties consented to the exchange, and whether they did so with adequate information about and understanding of what they were consenting to. In other words, justice in exchange demands informed consent.

The principle of informed consent has strong intellectual roots in biomedical ethics, where it has gained worldwide acceptance, as affirmed in a variety of national and international declarations in response to the Nazi medical atrocities revealed in the Nuremberg trials (Manson and O'Neill 2007). Subsequently becoming a standard of medical practice, the principle was introduced to avoid excessive paternalism, in particular situations where physicians took decisions on their patients' behalf, without confirming that these decisions were indeed in their patients' best interests. In the specific case of biomedical research, the principle aimed for participation to be achieved on a voluntary basis, with sufficient knowledge of the risk taken and the benefits of the research for society and one's community. Over recent decades this principle has been transferred to other fields: such as environmental policy, the provision of access to genetic resources, and technology assessment. In this chapter, when dealing with exchanges, free and informed consent has multiple requirements: people should be informed about any potential risks and benefits, they should be competent to assess and understand this information, and they must accept the object or procedure voluntarily (Shrader-Frechette 2002). This applies to both those acquiring and those selling objects of innovation.

First, people need full information on the risks and benefits. While documents about patented inventions are mostly freely accessible, access to critical studies is often difficult to obtain, especially for poorer parties, given that much of the academic literature is hidden behind paywalls. The information needs to be unbiased and show both the benefits and dangers of a technology, and provide sufficient information on alternatives. A government that is concerned about the safety of its citizens should support independent studies and make them freely available, as well as promoting scientific literacy and engaging in educational campaigns to improve risk awareness and knowledge of technological opportunities and limitations.

Second, there is the question of whether people, in granting consent, adequately understand the risks involved in both agreeing to and declining a procedure or object. The more we understand the complexities involved in interventions in our ecosystems, the harder it is to believe that we sufficiently understand the different factors at stake. This is so for both the anticipation of the effects of advanced biotechnology interventions (Kaebnick and Gusmano 2018) and the possession of extensive knowledge of the interacting organisms within agroecosystems (Gliessman 2007). It is a well-established fact that increasing technologization morally requires much more scientific and technological literacy and knowledge (Jonas 1979). Unfortunately, general scientific literacy does not increase at a rate that is proportional to the increase in the complexity of technologies and their effects on society, the environment, and interaction with other technologies (cf. Fernández Pinto 2019).

It is important for agriculture that both sellers and consumers are aware of the long-term effects of an invention, such as soil degradation and dependency on pesticides and fertilizers. Those selling the objects of innovation may have an interest in these innovations being used in accordance with specific instructions and in specific areas. For example, the emergence of pesticide resistance is also a problem for seed producers, and some have started to require the establishment of refuge areas, in order to reduce the speed of pesticide resistance by allowing room for natural predators and long-standing pathogen varieties (Timmermann 2015). Those buying the products need to know what relation of dependency they may end in up if they start to use them, even more so if they will have to use these products in higher doses in the future to maintain yields. A fully informed consumer should know about the potential risk factors, such as any effects on crop failure, or problems that may arise if she loses access to the product due to shortages, currency devaluation or poverty, and should have accurate knowledge of reasonable expectations (Thomas and De Tavernier 2017). Furthermore, users need to be aware of the costs of the skills they might end up losing, and the benefits of the new skills they will acquire by using the new products or processes. This is not an easy task due to the many hidden costs of conventional agriculture (Tittonell 2013).

Moreover, consent demands that people have agreed to a procedure while being fully capable of doing so. Here two differentiations are usually made. People cannot fully consent if they lack, momentarily or permanently, a sufficient mental capacity to analyse options, consequences, benefits and costs. For example—as sounds quite obvious to average citizens, but as remains widely ignored in practice—getting politicians, community leaders or business partners drunk or otherwise drugged in a "business dinner" reduces or even eliminates their capacity to consent, and provides moral grounds to invalidate any consent that is given.

Third, consent needs to be voluntary. There is a substantial difference in opinion on whether people can be under a significant amount of stress or pressure and still give consent that qualifies as legitimate. Forcing someone to consent under torture is the most obviously unacceptable case. Most types of threat or blackmail would also undermine the moral validity of consent. Yet there are some borderline cases that are difficult to assess. How should we evaluate a situation where a country threatens to cut development aid programmes for foreign governments or communities that are unwilling to give away highly needed biological samples, even when fairly remunerated? How should we assess business practices where a needed product is sold only when the buyer commits to buying or allowing the sales of other unwanted products, such as genetically modified seeds? It is a fact of life that we are continuously subject to certain pressures, as others may exert power over us through persuasion, seduction, manipulation or leadership (Brey 2008). Often, we have to choose the lesser of two evils: a difficult choice is not necessarily an imposed choice. Here a distinction can be drawn between cases where it is difficult to choose and cases where someone is forced to do something, which in a strict sense would require a significant amount of coercion (Vossenkuhl 2006). In general terms, it is a matter of thresholds whether we can recognize the validity of consent under pressure, thereby morally limiting the liberty to exert pressure, while making certain concessions to common causes and watching to not disproportionally burden specific groups.

As a matter of global justice, when assessing voluntariness, it could be argued that the background conditions for consent also need to be considered. People in low-lying Pacific islands may consent to pay a high price for climate change adaptation technologies that come from countries with historically high emissions rates, yet this does not immediately mean that such transactions are not morally questionable, as these people did not agreed to the background conditions that made such transactions necessary (Biddle 2016; Robaey and Timmermann 2018). While the decision to buy these adaptation technologies—for example, crops resistant to higher rates of salinity—may be voluntary, we may still oppose the profit-seeking commercialization of such technologies, as the background conditions that allow such exchanges are no accident.

A broader problem of securing consent is that of identifying the person or group who is morally entitled to grant consent. Particularly in the field of agriculture, many decision-making processes tend to be delegated, and we should ask ourselves how far this is legitimate. In agriculture, there is the additional problem that much innovation happens in a physical space that is not sufficiently controlled by those who develop inventions. At the farm level, large-scale land owners are often not the morally legitimate possessors of innovations developed on their property, such as improved seed varieties or locally-adapted farming methods. Moreover, the legal land owner often shows little respect for the intellectual interests of other people who live on the land, unless specific laws or social bonds are in play. Here we should not ignore the effects that local laws and customs have on people. In areas where women are disallowed or frequently impeded from legally acquiring and maintaining land (Nyéléni Forum for Food Sovereignty 2007; Llanque et al. 2018), this limits their ability to control any invention that they might develop on the land where they live. At the regional and state level, we may find that governments are often corrupt, or favour private or foreign interests at the expense of the interests of the local population (Escobar and Pardo 2007). Particularly in the case of landraces stewarded and developed by indigenous communities, the interests of national governments may be incompatible with those of indigenous people. Governments regularly ignore the interests of certain groups that they are supposed to protect. Racism, a colonial past, religious discrimination, anti-provincialism, gender biases, and sometimes mere geographic distance can all have a negative effect on protecting people's interests as buyers and sellers. In other cases, we find leaders within indigenous communities who have been reluctant to diffuse the traditional knowledge and varieties they possess, as this may weaken their power over the community (for Colombia, cf. Melo 2017). Deliberation and participatory decision-making processes often need to be established in order to obtain morally legitimate consent (Korthals 2008). This is particularly important when acquiring an invention based on traditional knowledge or involving genetic resources.

4.4 On Deceiving Transaction Partners

An issue closely related to informed consent is the problem of deceit. One of the biggest problems with agricultural innovation is that, generally, no substantial warranties are offered for the performance of innovations that involve biological materials. While a tractor sold to a farmer will be sold with a certain warranty over its functioning, this will not be the case for seeds. Furthermore, in the seeds trade, farmers will not know for certain what they have paid for until weeks or months after the seeds have sprouted, if they sprout at all. This opens the door to engaging in malpractice, particularly by making over-optimistic claims or, worse, by deliberately selling products that are unsuited to buyers' environmental and social conditions. The absence of warranties, and inefficient consumer protection institutions, make it particularly tempting to cease buying from official sales persons, and facilitate the existence of counterfeit seed markets. Such markets also harm consumers, as many seeds acquired through these channels do not contain their advertised traits, making it difficult for farmers to adjust inputs, manage pests and recover investments (Thomas and De Tavernier 2017).

Deceit also has a major impact on scientific research, which has the social role of generating realistic expectations and identifying potential hazards. We are currently witnessing the emergence of a large amount of historical research that proves the illicit influence of big companies on the research community: bribing and bullying scientists, and paying ghost-writers to publish favourable research findings and question unfavourable publications (Fernández Pinto 2015; di Norcia 2005). The food and agriculture sector has been particularly vulnerable to such practices, as exemplified by the key cases of industry denying the relations between sugar and caries (Kearns et al. 2015), between sugar-sweetened beverages and obesity (Marks 2019), and between tobacco and lung cancer, as well as denying the carcinogenicity of pesticides (Oreskes and Conway 2011). The borderline between what does and does not count as deceit is far from clear. For instance, a company has the right to reasonably question any evidence suggesting that their products are unsafe. Yet there is a difference between benefiting from a reasonable degree of doubt, and demanding a degree of scientific certainty and methodological accuracy that is prohibitively expensive or simply preposterous, especially when the possible hazards can be catastrophic, as still witnessed in the case of climate change denialism (Oreskes and Conway 2011). The current situation, where companies largely get to keep the profits they earned while casting doubt over harmful evidence, is particularly detrimental to society.

In a similar vein, it would also be deceptive to suggest that one has sufficient knowledge of the possible hazards and causal benefits of an innovation, when it is clear that one does not. This is even more worrisome when certain experiments are deliberately not carried out, or when lobbyists are hired to prevent potentially unfavourable research being done with taxpayers' money, so as to avoid market losses.

A particularly worrying form of deceit is to avoid telling users that the product on offer has not passed safety tests, or has even been banned in other countries. A product marketed as originating from a top laboratory in an industrialized country may give the impression that the product is, or at least can be, sold in the country of origin. There are many agrochemicals that continue to be sold in countries of the Global South despite being prohibited in the Global North due to their health and environmental hazards (Nasrala Neto et al. 2014; Atreya et al. 2011; Cabrera 2017).

4.5 Adequate Remuneration

The most challenging problem for applying principles of justice in exchange to agricultural innovation is to identify what amounts to adequate remuneration. There are multiple factors that make this particularly difficult. First, innovation, and especially innovation in agriculture, generally builds on the work done by an unidentifiable mass of other people. Second, inventions are generally only made possible by the existence and maintenance of social infrastructure and prior scientific work. Fairness would require that society's contributions towards these resources be discounted in the form of a social return (cf. Simon 2001) before researchers claim the fruits of their work. Third, the social interest of allowing people to benefit from the fruits of their own labour has to be weighed against other important social interests (Brody 1996), such as the human right to benefit from scientific advancement, and the human right to adequate food (Shaver 2010; De Schutter 2011). Fourth, intellectual property incentives do not adequately reflect fair rewards. Finally, a substantial amount, if not the majority of seed exchanges, as well as other objects of agricultural innovation, do not involve commercial transactions but rather transactions in kind, which bring social justice considerations of their own.

Let us start by discussing the first difficulty. The origins of agriculture date back to around 14,000 years ago, when people in the eastern Mediterranean started to select seeds from the best grasses in order to reseed them (McMichael 2017). More systematic efforts date back some 11,000 years (Mazoyer and Roudart 2006). Geographically, we can currently date back the origins of the most common crops to a number of centres, with the most important contributions to food security being made in the Euphrates region in the Middle East, Mesoamerica, the Andean region in and around Peru, the Horn of Africa, and East Asia, with smaller contributions from Papua New Guinea and the north-east of the United States. Substantial efforts have been made to improve domesticated crops brought from other areas. The Columbian Exchange, referring to the introduction of wheat, olives and grapes to the Americas and the adoption of potato, maize and tomato varieties in Europe, illustrates the extent of this exchange (Kloppenburg 2005). It would be ridiculous to undervalue the importance of the innovations in imported crop varieties yielded by this exchange for agriculture and cuisine on both sides of the Atlantic, and the adoption of crops from other parts of the world. World cuisine would otherwise be much impoverished: imagine Italian dishes without the introduction of tomatoes from the Americas, or the absence of plantains, originating from South-East Asia, in Caribbean meals. These crop exchanges and importations date back far enough for the innovative value added to these crops outside their centres of origin to be undeniable, and we should also recognize the effort involved in conserving these resources, which are sometimes even found only outside their centres of origin. This posterior work deserves both compensation and recognition.

With respect to the development of agricultural methods and improvements, it is even more difficult to assess the value of prior contributions. Ecological knowledge of plants and their dietary effects on humans has been acquired through centuries. The same goes for methods of propagating plants, managing pests, improving soil fertility, administering water, and fixing damaged plants. These methods and techniques have been developed, improved and tested over countless generations and adapted to a wide variety of environmental and social conditions (Reyes-García et al. 2018). It would be senseless to weigh the effort involved in one's own contribution against the historical work underlying these innovations.

As a second issue, society spends a substantial amount of resources on stimulating and facilitating innovation (Stephan 2012), which creates an obligation to give back to society, in the form of a social return. Much innovation does not happen in a social vacuum. Experiments and research that are carried out over longer of periods of time or include a larger number of participants require a certain amount of stability, which is usually achieved through social cooperation and the maintenance of infrastructure, which—depending on the research being carried out—might be quite significant, as is the case with dependency on water and electricity supply and global communication networks. While some may claim that one discharges one's duty towards the maintenance of these common goods by paying fixed contributions in the form of taxes, others claim that one should contribute back to such common goods in proportion to the degree that one has benefited from them, as these common goods not only make it easier to do research, but are indispensable for such endeavours. Which of these sides is right is hard to settle. At least when it comes to taxation, we can observe that most countries have embraced progressive taxation systems: something that reveals a widespread acceptance of the principle that people who benefit more from social cooperation schemes should also pay more towards their maintenance.

The third problem concerns the balancing of public and private interests. Property rights are not absolute, and securing these rights cannot be done without taking account of their effect on securing human rights, especially when basic rights such as the right to adequate food are at stake (Marks 2011; Timmermann 2014). Two major interests are involved: an interest in the material benefits of one's labour, and an interest in benefiting from scientific advancement, which requires access to innovation. Both of these interests are protected by the Universal Declaration of Human Rights.[1]

To truly serve private economic interests, intellectual property holders need to exclude those people who are able to contribute towards the costs of production but are reluctant to do so, unless obliged. To maximize income, there is a major problem with pricing. There will always be people who are willing to contribute towards the costs of production, but are only able to pay somewhat less than the requested

[1]Universal Declaration of Human Rights (1948), art. 27.

price (Timmermann 2017). This may lead to lost income, and has substantial costs in terms of welfare. Moreover, some people have no disposable income at all, and are therefore unable to contribute towards the satisfaction of innovators' economic interests. This again leads to wastage, in terms of lost welfare. Many people are left empty-handed as a consequence of our global innovation reward system.

Moreover, the extreme levels of inequality found in the world make it difficult to argue that innovators' legitimate expectations should be always met. When individuals engage in activities that are encouraged by existing institutions, one could in principle argue that those people have legitimate expectations of receiving the legal entitlements that result from such activities (Hsieh 2000). People who engage, however, in inventive activities to develop marketable products for food production operate under background conditions that are far from ideal, and some are extremely unjust. Addressing innovators' legitimate financial expectations cannot justify blocking access to a vital good that could be made more widely available without inflicting any financial cost. Innovators have a moral obligation to refrain from engaging in the socially and environmentally unsustainable exploitation of their inventions, even when this comes at a financial disadvantage. A balance needs to be struck between providing sufficient financial incentives and opportunities to recoup costs, so as to sustain the socially desirable activity of innovating, while ensuring that such remunerations do not jeopardize the capacity to sustainably provide access to the objects of innovation.

The link between access to innovation and securing basic needs is a complex one and difficult to assess from an ethical perspective. As a first issue, there is a collective action problem. Securing the right to food depends on a series of factors, among which inventions play a significant role. If a single innovator withholds her innovation, this will rarely affect food security. However, if many innovators withhold their improvements, we will see a negative influence on harvest yields. The same goes for abusive prices, which reduce access and diffusion, and therefore reduce the impact of innovation.

There is also the problem of sharing social burdens fairly. If we decide to burden agricultural research and development with additional tasks that other sectors do not have to assume, we may create a disincentive to work in these sectors in the first place, and an issue of justice by taxing a certain line of work more than other lines. It is clearly unacceptable that lines of work which yield specific benefits for society should face additional burdens, while those engaged in leisure or luxury activities enjoy far more liberties. This could encourage researchers to simply switch to areas of work which do not serve a social need.

Also of special importance is the fact that a substantial amount of research and development is done or partly financed by private parties. If private companies are the first to patent, this may hinder public institutions from offering a substitute when a similar research line was pursued, or when genetic resources were involved (Schroeder and Pogge 2009). There are, however, plenty of cases where the private sector is the only one working in an area where there is a strong need. This leaves users with a choice between attempting to acquire an expensive product offered by the private sector or having no solution at all. While there is a lot of opportunistic behaviour among private companies, there is also substantial corruption and slack in the public sector. Taxpayers' research money regularly ends up in inefficient institutions, where researchers favour personal interests, rather than setting up research agendas and hiring the most suitable candidates so as to optimally advance science. It is crucial that we improve good scientific and professional practice in public institutions all over the world, as market incentives rarely play a role in punishing inefficient public institutions.

A fourth problem is that intellectual property inadequately reflects effort and social need. The reward that one may receive for putting in a huge amount of effort will not necessarily be higher than the reward received for an invention that benefited from a substantial amount of luck. The effort put into an invention bears no relation to the size of the financial reward. Even if we take into account that being able to perceive certain opportunities may involve substantial training and experience (Cwik 2014), which need to be compensated, reliance on intellectual property rights is a bad proxy for identifying fair rewards. A product sought by rich customers is likely to receive higher rewards than one that is desperately needed by poorer customers (Hollis and Pogge 2008). The size of the reward is also highly dependent on factors that innovators and their investors did not bring about, and which therefore cannot justify desert, such as the existence of competing products, social and environmental changes, access to credits, and other entry barriers (Christman 1988). It is also unclear why innovators should be entitled to benefit from intellectual property rights beyond their borders (James 2012). On the other hand, if society places a special value on rewarding effort, then failed inventions and independent late inventions should also be rewarded: something which is not done by intellectual property regimes (Sterckx 2006). Most importantly, much innovation in agriculture does not even qualify for intellectual property rights, and is often done by people who lack the economic means and expertise to apply for such rights (De Jonge 2011).

Another problem is temporary exclusivity in intellectual property. The setting of prices allows some to have immediate access, while others will have to wait another decade or more, which may have substantial justice implications for vital products. There needs to be a balance between allowing innovators to seek financial benefits and providing timely access (James 2012). Deferred access comes too late for those already dead and is too burdensome for those who have to endure the substantial suffering, crop failures and hunger that are involved.

Enormous power imbalances between buyers and sellers permit exploitation—understood as unilaterally enforcing disadvantageous exchange conditions that would never have been accepted in a level bargaining situation—which can manifest itself in different forms. There are hard cases, where people deserve benefits but receive nothing, other cases where the benefit they receive is clearly insufficient for the services they have provided, and also cases where the benefits people receive meet their own expectations, but are clearly unfair when objectively analysed (Arnason and Schroeder 2013). This last case is difficult to solve, because of the problem of adaptive preferences, which is often manifest in historically discriminated communities who may not have high expectations when it comes to obtaining a fair reward.

Lastly, the fact that an exchange was carried out without relying on money does not make it irrelevant for justice in exchange. Informal seed exchanges still dominate in the world (Jackson et al. 2007, De Jonge et al. 2015). The fact that someone asks for seeds in return does not make a trade less exploitative, particularly when we are not talking about a one-to-one exchange ratio. Similarly, making someone work for an excessive number of hours on one's farm in exchange for being taught a particular agricultural technique, despite saving future work and being genuinely innovative, may still amount to exploitation if positions of vulnerability are used to extract disproportional benefits. In times and areas of weak labour protection laws, we should be particularly critical of exploitative labour arrangements masked as apprenticeship or training.

4.6 Avoiding Harm

Generally speaking, justice in exchange requires that transactions do not lead to harm. As mentioned before, this requirement can be understood in a narrow or a broad sense. To avoid harm in a transaction between two parties, vulnerabilities among buyers and sellers should not be exploited, people should not end up in a position of harmful dependency, and the invention should have no harmful effect on the user or her property. In a broader sense, commercialization can lead to indirect harm, especially by abruptly changing the market demand of a product, or of resources upon which people's survival or traditional practices depend. Let us analyse these cases.

The most obvious case of harm is the reckless exploitation of vulnerabilities. The strong link between agricultural innovation and food security may leave commercial transactions at particular risk of exploitation. Farmers suffering the consequences of droughts or pests, or fearing these consequences, can be easy targets for the sale of overpriced or inadequate technological solutions, sometimes even including agrochemicals that are strictly prohibited in other countries after major health hazards were revealed. Particularly when financial credits are offered without proper control or ethical responsibility, farmers may end up facing major financial struggles, which sometimes even lead to suicide (Thomas and De Tavernier 2017).

Different types of actors are vulnerable to exploitation. Corporations may also exploit situations of vulnerability to achieve highly unfair access and benefit agreements in order to get access to genetic resources. Owners of larger farms may in times of financial crisis, on the other hand, exploit smaller companies who offer their technological innovations.

Scholars who work on access and benefit-sharing agreements have pointed out that those who contribute to science and technology should be able to access the fruits of the research they help to develop, for otherwise we may encounter exploitative arrangements (Arnason and Schroeder 2013). While this idea is relatively new in agricultural research, it is theoretically based on a well-anchored principle in biomedical research, applied particularly to research concerning human subjects, which condemns participation when it involves risks and does not benefit participants or their community (Wenner 2015).

The acquisition of new technologies, or the reliance on a few customers as a service provider, may also lead to a position of dependency that leaves people extremely vulnerable to exploitation and losses of harvest. Farmers who decide to switch to high-input conventional agriculture may underestimate the dangers of recessions and currency devaluations to their ability to acquire the needed inputs in the future. Over a two-year period, sudden losses of currency values of 25–30% in relation to major currencies, such as the Euro or the Dollar, are common in low- and middle-income countries: a factor that may hinder the acquisition of foreign agricultural inputs. Farmers may also be unaware of the difficulty involved in switching back to ecological intensification, especially when traditional seed varieties are lost, and skills are forgotten or eroded. One case study that has received ample attention is Cuba, where the food systems had to be redesigned for self-sufficiency, after an embargo by the United States made it extremely difficult and costly for farmers to acquire agricultural inputs, tools and replacement parts for machinery (Rosset et al. 2011; Funes-Monzote 2008).

Farmers who use certain technologies need to be fully aware of both their risks and their benefits. More awareness is needed of the harmful effects of agrochemicals on waterways and soil bacteria. Here it is crucial that farmers factor into the acquisition of new technologies any new costs which their use may cause. If farmers or their neighbours will have to rely on bottled water or buy expensive water purifiers because of the contamination of waterways by agrochemicals, the use of such inputs may end up being unprofitable; or may unfairly impose costs on third parties, which in turn may trigger retaliation from neighbours, ultimately inflicting additional costs on the farmers who use such agrochemicals. Agrochemicals and heavy machinery may also destroy organisms that have been providing valuable ecosystem services for farmers.

Similarly, there is also the issue of how the acquisition of genetically modified seeds affects farmers and their neighbours. Seeds, unlike other inventions, multiply themselves. Current legal practices are unfairly beneficial to the seed industry, as they make farmers liable for patent infringements if seeds end up growing in their fields, as shown by the case of *Bowman v. Monsanto Co.* (Robaey 2016). This is also a major issue for neighbouring organic farmers, who may lose their organic certification when pollen from other farmers contaminates their fields. In addition, gene flows from genetically modified crops can in certain cases change wild relatives with negative consequences for biodiversity and future food security (Ervin et al. 2010). Widespread use of pesticides also harms bee colonies, ruining the livelihood of many apiculturists.

Concerning indirect harm, the large-scale commercial exploitation of indigenous crops may have a negative effect on the overall accessibility of these crops in their centres of origin (Soria-López and Fuentes-Páramo 2016). Such cases are particularly controversial when biopiracy—or benefits being insufficiently shared—leads to the massive exploitation of a genetic resource, leaving the people who have traditionally stewarded and benefited from this good empty-handed, due to price increases or unsustainable extraction.

Another issue is the effect of massive sales. While, from an economic perspective, achieving complete market dominance is the most ambitious goal, this would be ecologically and socially disastrous if pursued by many agricultural innovators. Being able to combat a pest with only one major pesticide would make farmers extremely vulnerable once the organisms develop resistance. Massive sales of blockbuster seeds lead to enormous monocultures: something that can cause serious environmental problems and is a major risk for food security. To maintain agrobiodiversity, seed monocultures—that is, the massive sales of specific seeds—need to be controlled, even if there is a high demand for more of the same seeds. Vast areas ignore such precautions, as shown by the immensity of monoculture soy fields in South America (Leguizamón 2016). The pursuit of private interests should not lead to public harms such as genetic erosion and public health hazards (Arancibia et al. 2020). To maintain agrobiodiversity, policies need to be established to re-establish and maintain this good as a public good (Timmermann and Robaey 2016).

To avoid harm, regulation is often seen as the answer. When enacting regulations, it must be kept in mind that we live in a globalized world with common problems and concerns. Countries that compete against each other for foreign investments may be obliged to undergo an environmental and social "race-to-the-bottom": offering lower and lower standards to increase competitiveness (Hassoun 2009; Ngosso 2013). We can also observe such tendencies within countries (Sklair 2016). Here regulations need to be established and enforced at a national and global level, to avoid the result that companies move to regions and countries with laxer legal protections, while countries desperate for foreign investments lower their standards to attract even companies that ruthlessly seek to maximize profits.

Theories of competitive markets generally assume that people seek to maximize their income and do not use force, power or collusion to improve their bargaining power and market position (Koslowski 2008). Here too we see a number of violations of free market principles. Commercial interactions are often not entered into freely. Major corporations, funding agencies and donors often use their power to impose certain agricultural models and technologies over others (Patel 2013). Food processors and retailers have colluded to oppose food labelling (Pollan 2008). Lobbyists have pushed for laws requiring the use of certified seeds for food production, so as to thereby limit the use and exchange of freely available landraces. It is in the long-term interests of the companies who benefit from the market system to support justice in exchange, and not to free-ride on efforts to improve trust and reliance on markets (Koslowski 2008). Distrust in markets invites people to be self-reliant, and impedes benefiting from the division of labour and specialization, countering the very justification for markets.

4.7 Recognition

A question that has recently been introduced into the debate is how far we owe transaction partners recognition. Besky (2015), for example, discusses the issue of recognition as an additional factor that needs to be considered as part of fair trade, beside other well-established demands such as fair wages and decent working conditions. Recognition requires visibility, and acknowledgement for the effort made to secure something of importance: in our case, the optimization of food production. We can see this call for recognition among the demands raised by the food sovereignty movement, which has strongly criticized how the contributions of smallholders—and here particularly of women—for food security are downplayed (Nyéléni Forum for Food Sovereignty 2007). This demand for recognition reveals that exchanges are not only meant to satisfy economic interests, but also immaterial interests: the recognition of one's skills, individuality and social contribution, independently of whether one is in a position of dependency or mutual influence. Recognition for providing valuable goods and ecological services can play a crucial role in maintaining interest in working in areas that are underpaid or unpaid, as is the case with much work in the optimization of food production and processing (Timmermann and Félix 2015). We will return to the issue of recognition later, in the chapter on contributive justice.

4.8 Foundational Question: On Having a Market at All

An emphasis on the importance of justice in exchange should not imply that delivering agricultural innovation through market systems is unavoidable or generally recommended. We are free to question the very use of market incentives. To start, one problem of seeking compensation in a world of extreme inequality is that some people who need access to a good will end up empty-handed, because they will not be able to pay even a reduced price for the good in question (Timmermann 2017). The idea of needing to recoup research and development costs by selling vital innovations is in itself problematic, since people will be denied access even when they are unable to contribute towards the cost of production, and it does not cost any additional resources to allow them access on a humanitarian basis (Wilson 2007). This outcome has substantial moral weight when it deprives people from access to innovations that could reduce hunger and malnutrition. Market incentives are not indispensable for stimulating innovation. As we have seen, especially in relation to access to medicines, there are a number of well-elaborated alternative proposals for making medicines accessible and available, without relying on high mark-up prices to finance research and development (Outterson 2014; Hollis and Pogge 2008). Furthermore, new information technologies and new types of licensing schemes (e.g. creative commons) facilitate new types of collaborative research and diffusion of innovations, which need not involve market transactions (Benkler 2017). There are currently a number of initiatives for considering food as a commons, to make sure everyone has access to it, as well as having a say over its production (Vivero-Pol et al. 2018; Ferrando et al. forthcoming). Another alternative is to provide for agricultural innovation as a public good, in the sense that nobody should be excluded from accessing the inventions they need. Many public institutions and international organizations have provided and still provide improved seed varieties, optimized irrigation systems and pest management methods free of charge. We can strengthen these institutions so that they better serve farmers.

Efforts to decommodify agricultural innovation have multiple advantages. For instance, as seen in the previous chapter, whenever there is a technological solution in the public domain, competing commercial products will have to offer added advantages in order to find buyers, by providing additional customer services, features or effectiveness. This will increase competition to the benefit of the public. There are also social advantages. Maintaining agricultural commons can create a sense of community and provides a sphere where social relationships can be developed for mutual benefit (Vail 2010). Common spaces allowing user innovation facilitate the development of key ethical virtues, such as sociability, camaraderie, friendship and cooperation (Benkler and Nissenbaum 2006). Once sharing on a reciprocal basis is (re-)established, this provides a foundation for other types of cooperative problem-solving endeavours.

Many types of commons have been active since time immemorial, while others have been developed only recently. Traditional seed exchange practices are still dominant, and serve as a major means for the exchange of knowledge and traditions, contributing to social cohesion between different groups (Montenegro de Wit 2019). Seeds that have been stewarded and improved are often also a major source of pride and recognition for farmers (Frison and Coolsaet 2018). There are also land commons arrangements that allow the free exchange of observations, experiences and innovations that are specifically suited to local social and environmental conditions, and that can be shown as examples to other communities (Maughan and Ferrando 2018).

Enthusiasm for providing innovation on a non-commercial basis should not blind us to critically assessing any possible ethical issues. A mandate to avoid harm is also relevant for exchanges that do not involve money. The acquisition of innovations through informal transactions or open repositories still has to be treated with caution, as often nobody has assumed responsibility for the possible risks such innovations may pose (Gupta et al. 2016). Such dangers are also presented in ecological agriculture, as shown by the well-known dangers of wrongly handling human faeces as fertilizers (Jenkins 2005), and the husbandry of animals in close proximity to people. The absence of financial compensation does not cancel obligations concerning informed consent. There is also an issue of fairness, if society free-rides on the vital creative efforts of some without providing anything—from infrastructure to formal acknowledgement—in return.

In some cases, mixing open science with the provision of commercial services has been fruitful. Open-source software shows that inclusive development programmes are compatible with business models that offer added services on a commercial basis. We need not choose between market and non-market models: mixtures are possible (Benkler 2017).

4.9 Feasibility: Achieving Justice in Exchange

It is practically impossible to adequately reward all those who have directly or indirectly contributed towards an invention. Innovators are inspired by a wide range of sources during their creative processes, and the boundary between those who deserve compensation and recognition for their contributions and those who do not—be it moral or economic—is far from clear, as organizers of large-scale citizen science projects in particular have come to realize (Riesch and Potter 2014). With respect to access to the objects of innovation, the strong link between agricultural innovation and the human right to adequate food obliges the design of sales strategies that avoid (or reduce to a minimum) the outcome of people in need being left empty-handed. Due to the extreme inequalities that can be observed between and within countries, this may require the subvention of products for poorer population groups. Since food security is a global public good, the cost of providing this good needs to be shared among all people worldwide (Timmermann 2018).

Yet the major problems of access to innovation and providing a fair reward should not keep us from judging whether the wider set of demands of justice in exchange (Table 4.1) are being met.

Adopting this broader perspective, we must be fully aware that transactions that ignore the demands of justice in exchange are likely to have harmful effects on social cohesion, as people are thus not perceived as equals and such agreements may lead to resentment and objections to exploring further collaborative opportunities (Cozzens 2007). Maintaining willingness to cooperate has multiple advantages, as trust in market systems allows economies of scale and specialization, and facilitates the production of surpluses that can be used to mitigate risks (Heath 2006). In this sense, trust in markets has a public good character, as everyone who relies on markets

Table 4.1 Elements of justice in exchange and their difficulties

Element	Difficulties
Entitlement	Identifying the legitimate person or institution to authorize exchanges
Informed consent	Degree of scientific, political and legal literacy and adequate understanding
Deceit	Possible different interpretations, difficulty of assessing intention
Remuneration	Objective and subjective adequacy of remuneration may differ
No harm	No clear consensus on the limits of liability
Recognition	Different expectations

benefits from their advantages, but all suffer (albeit in varying degrees) when individual actors misuse this trust for their profit-seeking. Unfortunately, there is a limit on how much fairness in exchange can be achieved in practice. There is also a price to be paid in temporarily withholding access while seeking that the various demands of justice in exchange be met. Especially when larger innovators do not comply with good practice, it may be very costly to boycott them, as they provide an essential good. Untimely access may negatively affect food production. In real-word circumstances, we may need to make some compromises concerning what ideals would require, in order to ensure timely access.

There are, however, options for giving farmers more power to seek that irresponsible innovators are punished. Governments could be more tolerant in treating large-scale refusal to pay licensing fees for seeds as a protest against companies who engage in exploitative practices. In addition, farming schools that teach agroecological methods that allow increased self-sufficiency should be expanded, so that farmers can produce their own inputs when markets do not provide inputs on fair terms. Moreover, institutions that protect consumers and watch out for unfair competition should be strengthened.

For justice in exchange, background conditions also play a major role. The context in which a certain type of traditional knowledge was appropriated matters (Schroeder 2009). It seems that something makes cases morally different, depending on who appropriates something from whom. Writing from a Swiss context, Anna Deplazes-Zemp (2018) ponders the lack of public outcry from traditional alpine communities about the appropriation by herb candy producers of traditional herbal knowledge about coughs, in contrast with cases where knowledge of this type was appropriated from the Global South. Perhaps traditional alpine communities feel that they have benefited much more from sharing (willingly or unwillingly) with the wider society in which they are embedded, in comparison with traditional communities from formerly colonized nations. Another possible answer is that herb candy producers are much more likely to advertise that they are using herbal knowledge from a specific geographic area, thereby being open about the intellectual origin of their products and thus providing recognition. We will return to this point later, in the chapter on restorative justice, as such contextualization is difficult to fully assess without taking historical injustices into consideration.

The extreme deprivation in which many smallholders around the world find themselves obliges us to strengthen public institutions, in order to provide innovations that will improve their food production capabilities and reduce the hardship of agricultural work. We can only rely on market incentives for innovation in cases where people have disposable income. When this is not the case, as with the nearly half of the world population who live on less than two and a half dollars a day, a stubborn reliance on markets to diffuse innovations will not allow people to benefit from scientific advancement: a too often neglected human right. Strong public research institutions should provide the bulk of needed innovations, and a prize system should be established to reward innovative farmers, formally recognize their creativity, and encourage further innovation.

References

Arancibia, Florencia, Renata Campos Motta, and Peter Clausing. 2020. The neglected burden of agricultural intensification: A contribution to the debate on land-use change. *Journal of Land Use Science* 15 (2–3): 235–251. https://doi.org/10.1080/1747423X.2019.1659431.

Arnason, Gardar, and Doris Schroeder. 2013. Exploring central philosophical concepts in benefit sharing: Vulnerability, exploitation and undue inducement. In *Benefit sharing*, ed. by Doris Schroeder and Julie Cook Lucas, 9–31. Dordrecht: Springer.

Atreya, Kishor, Bishal K. Sitaula, Fred H. Johnsen, and Roshan M. Bajracharya. 2011. Continuing issues in the limitations of pesticide use in developing countries. *Journal of Agricultural and Environmental Ethics* 24 (1): 49–62.

Benkler, Yochai. 2017. Peer production, the commons, and the future of the firm. *Strategic Organization* 15 (2): 264–274.

Benkler, Yochai, and Helen Nissenbaum. 2006. Commons-based peer production and virtue. *The Journal of Political Philosophy* 14 (4): 394–419.

Besky, Sarah. 2015. Agricultural justice, abnormal justice? An analysis of fair trade's plantation problem. *Antipode* 47 (5): 1141–1160.

Biddle, Justin B. 2016. Intellectual property rights and global climate change: Toward resolving an apparent dilemma. *Ethics, Policy & Environment* 19 (3): 301–319.

Brey, Philip. 2008. The technological construction of social power. *Social Epistemology* 22 (1): 71–95.

Brody, Baruch. 1996. Public goods and fair prices: Balancing technological innovation with social well-being. *The Hastings Center Report* 26 (2): 5–11.

Cabrera, Laura Y. 2017. Pesticides: A case domain for environmental neuroethics. *Cambridge Quarterly of Healthcare Ethics* 26 (4): 602–615.

Christman, John. 1988. Entrepreneurs, profits, and deserving market shares. *Social Philosophy and Policy* 6 (1): 1–16.

Cozzens, Susan E. 2007. Distributive justice in science and technology policy. *Science and Public Policy* 34 (2): 85–94.

Cwik, Bryan. 2014. Labor as the basis for intellectual property rights. *Ethical Theory and Moral Practice* 17 (4): 681–695.

De Jonge, Bram. 2011. What is Fair and Equitable Benefit-sharing? *Journal of Agricultural and Environmental Ethics* 24 (2): 127–146.

De Jonge, Bram, Niels P Louwaars, and Julian Kinderlerer. 2015. A solution to the controversy on plant variety protection in Africa. *Nature Biotechnology* 33 (5): 487–488

De Schutter, Olivier. 2011. The right of everyone to enjoy the benefits from scientific progress and the right to food: From conflict to complementarity. *Human Rights Quarterly* 33: 304–350.

Deplazes-Zemp, Anna. 2018. Commutative justice and access and benefit sharing for genetic resources. *Ethics, Policy & Environment* 21 (1): 110–126.

di Norcia, Vincent. 2005. Intellectual property and the commercialization of research and development. *Science and engineering ethics* 11 (2): 203–219.

Dieterlen, Paulette. 2003. *La pobreza: Un estudio filosófico*. México, DF: Fondo de Cultura Económica.

Dodds, Walter K. 2005. The commons, game theory and aspects of human nature that may allow conservation of global resources. *Environmental Values* 14: 411–425. https://doi.org/10.3197/096327105774462683.

Ervin, David E., Leland L. Glenna, and Raymond A. Jussaume. 2010. Are biotechnology and sustainable agriculture compatible? *Renewable Agriculture and Food Systems* 25 (2): 143–157.

Escobar, Arturo, and Mauricio Pardo. 2007. Social Movements and biodiversity on the Pacific Coast of Colombia. In *Another world is possible: Beyond Northern epistemologies*, edited by Boaventura de Sousa Santos, 288–314. London: Verso.

ETC Group. 2008. *Who owns nature?: Corporate power and the final frontier in the commodification of life*. Ottawa: ETC Group.

Fernández Pinto, Manuela. 2015. Commercialization and the limits of well-ordered science. *Perspectives on Science* 23 (2): 173–191.

Fernández Pinto, Manuela. 2019. Scientific ignorance. *Theoria: An International Journal for Theory, History and Foundations of Science* 34 (2):195–211.

Ferrando, Tomaso, Priscilla Claeys, Dagmar Diesner, Jose Vivero Pol, and Deirdre Woods. Forthcoming. Commons and commoning for a just agroecological transition: How to decolonise and decommodify our food systems. In *Resourcing an agroecological urbanism: political, transformational and territorial dimensions*. London: Routledge.

Frison, Christine, and Brendan Coolsaet. 2018. Routledge handbook of food as a commons. In *Routledge Handbook of Food as a Commons*, ed. by José Luis Vivero-Pol, Tomasso Ferrando, Olivier De Schutter, and Ugo Mattei, 218–230. Oxon & New York: Routledge.

Funes-Monzote, Fernando. 2008. Farming like we're here to stay: The mixed farming alternative for Cuba. Ph.D. diss., Wageningen University.

Gliessman, Stephen R. 2007. *Agroecology: The ecology of sustainable food systems*. Boca Raton: CRC Press.

Gupta, Anil K., Anamika R Dey, Chintan Shinde, Hiranmay Mahanta, Chetan Patel, Ramesh Patel, Nirmal Sahay, Balram Sahu, P Vivekanandan, and Sundaram Verma. 2016. Theory of open inclusive innovation for reciprocal, responsive and respectful outcomes: coping creatively with climatic and institutional risks. *Journal of Open Innovation: Technology, Market, and Complexity* 2 (1): 16.

Hassoun, Nicole. 2009. Free trade and the environment. *Environmental Ethics* 31 (1): 51–66.

Heath, Joseph. 2006. The benefits of cooperation. *Philosophy & Public Affairs* 34 (4): 313–351. https://doi.org/10.1111/j.1088-4963.2006.00073.x.

Heller, Michael. 2008. *The gridlock economy: How too much ownership wrecks markets, stops innovation, and costs lives*. New York: Basic Books.

Hollis, Aidan, and Thomas W. Pogge. 2008. The health impact fund: Making new medicines accessible for All. Oslo & New Haven: Incentives for Global Health. https://www.healthimpactfund.org/hif_book.pdf. Accessed 31 Mar 2012.

Hsieh, Nien-hê. 2000. Moral desert, fairness and legitimate expectations in the market. *Journal of Political Philosophy* 8 (1): 91–114.

Jackson, Louise E., U. Pascual, and T. Hodgkin. 2007. Utilizing and conserving agrobiodiversity in agricultural landscapes. *Agriculture, ecosystems & environment* 121 (3): 196–210.

James, Aaron. 2012. *Fairness in practice: A social contract for a global economy*. New York: Oxford University Press.

Jenkins, Joseph C. 2005. *The humanure handbook: A guide to composting human manure*, 3rd ed. Jenkins Pub.

Jonas, Hans. 1979. *Das Prinzip Verantwortung: Versuch einer Ethik für die technologische Zivilisation*. Frankfurt am Main: Insel-Verlag.

Kaebnick, Gregory E., and Michael K. Gusmano. 2018. Making policies about emerging technologies. *Hastings Center Report* 48: S2–S11.

Kearns, Cristin E., Stanton A. Glantz, and Laura A. Schmidt. 2015. Sugar industry influence on the scientific agenda of the National Institute of Dental Research's 1971 National Caries Program: A historical analysis of internal documents. *PLoS medicine* 12 (3): e1001798.

Kloppenburg, Jack. 2005. *First the seed: The political economy of plant biotechnology*, 2nd ed. Madison: University of Wisconsin Press.

Koller, Peter. 2013. Social and global justice. In *Spheres of global justice: Volume 1 Global challenges to liberal democracy. Political participation, minorities and migrations*, ed. by Jean-Christophe Merle, 433–443. Dordrecht: Springer.

Korthals, Michiel. 2008. Ethics and politics of food: Toward a deliberative perspective. *Journal of Social Philosophy* 39 (3): 445–463.

Koslowski, Peter. 2008. Some principles of ethical economy. In *Trends in business and economic ethics*, ed. by Christopher Cowton and Michaela Haase, 31–70. Berlin & Heidelberg: Springer.

Leguizamón, Amalia. 2016. Disappearing nature? Agribusiness, biotechnology and distance in Argentine soybean production. *The Journal of Peasant Studies* 43 (2): 313–330.

Llanque, Aymara, Ana Dorrego, Giulia Costanzo, Bishelly Elías, and Georgina Catacora-Vargas. 2018. Mujeres, trabajo de cuidado y agroecología: Hacia la sustentabilidad de la vida a partir de experiencias en diferentes eco-regiones de Bolivia. In *Agroecología en Femenino: Reflexiones a partir de nuestras experiencias*, ed. Gloria Patricia Zuluaga. Sánchez, Georgina Catacora-Vargas, and Emma Siliprandi, 123–139. La Paz: SOCLA.

Manson, Neil C., and Onora O'Neill. 2007. *Rethinking informed consent in bioethics*. Cambridge: Cambridge University Press.

Marks, Stephen P. 2011. The neglected human right to benefit from scientific progress: implications for human development. In *Human Development and Capabilities Association 2011 International Conference*, Den Haag, 6–8 September 2011.

Marks, Jonathan H. 2019. *The perils of partnership: Industry influence, institutional integrity, and public health*. New York: Oxford University Press.

Maughan, Chris, and Tomaso Ferrando. 2018. Land as a commons. In *Routledge handbook of food as a commons*, ed. by José Luis Vivero-Pol, Tomasso Ferrando, Olivier De Schutter and Ugo Mattei. Oxon & New York: Routledge.

May, Christopher. 2006. Patents, universities and the provision of social goods in the information society. *Ethical Perspectives* 13 (2): 289–304.

Mazoyer, Marcel, and Laurence Roudart. 2006. *A history of world agriculture: From the neolithic age to the current crisis*. New York: Monthy Review Press.

McMichael, Anthony. 2017. *Climate change and the health of nations: Famines, fevers, and the fate of populations*. Oxford: Oxford University Press.

Melo, Jorge Orlando. 2017. *Historia mínima de Colombia*. Ciudad de México: El Colegio de México.

Montenegro de Wit, Maywa. 2019. Beating the bounds: How does 'open source' become a seed commons? *The Journal of Peasant Studies* 46 (1): 44–79.

Nasrala Neto, Elias, Francisco Antonio de Castro Lacaz, and Wanderlei Antonio Pignati. 2014. Health surveillance and agribusiness: The impact of pesticides on health and the environment. Danger ahead! *Ciencia & saude coletiva* 19: 4709–4718.

Ngosso, Thierry. 2013. The right to development of developing countries: An argument against environmental protection? *Public Reason* 5 (2): 3–21.

Nyéléni Forum for Food Sovereignty. 2007. *Declaration of Nyéléni*. Sélingue: Nyéléni Forum for Food Sovereignty.

Ooms, Gorik. 2010. Why the West is perceived as being unworthy of cooperation. *Journal of Law, Medicine and Ethics* 38 (3): 594–613. https://doi.org/10.1111/j.1748-720X.2010.00514.x.

Oreskes, Naomi, and Erik M. Conway. 2011. *Merchants of doubt: How a handful of scientists obscured the truth on issues from tobacco smoke to global warming*. New York: Bloomsbury.

Outterson, Kevin. 2014. New business models for sustainable antibiotics. In *Centre on global health security working group papers, working groups on antimicrobial resistance, Paper 1*. London: Chatham House (The Royal Institute of International Affairs).

Patel, Raj. 2013. The long green revolution. *The Journal of Peasant Studies* 40 (1): 1–63.

Patton, Paul. 2005. Historic injustice and the possibility of supersession. *Journal of intercultural studies* 26 (3): 255–266.

Pogge, Thomas W. 2008. *World poverty and human rights: Cosmopolitan responsibilities and reforms*, 2nd ed. Cambridge: Polity.

Pollan, Michael. 2008. *In defense of food: An eater's manifesto*. New York: Penguin.

Reyes-García, Victoria, Petra Benyei, and Laura Calvet-Mir. 2018. Traditional Agricultural Knowledge as a Commons. In *Routledge handbook of food as a commons*, ed. by José Luis Vivero-Pol, Tomasso Ferrando, Olivier De Schutter, and Ugo Mattei, 173–184. Oxon & New York: Routledge.

Riesch, Hauke, and Clive Potter. 2014. Citizen science as seen by scientists: Methodological, epistemological and ethical dimensions. *Public understanding of science* 23 (1): 107–120.

Robaey, Zoë. 2016. Gone with the wind: Conceiving of moral responsibility in the case of GMO contamination. *Science and engineering ethics* 22 (3): 889–906.

Robaey, Zoë, and Cristian Timmermann. 2018. Fair agricultural innovation for a changing climate. In *Food justice, the environment, and climate change*, ed. by Erinn Cunniff Gilson and Sarah Kenehan, 213–230. Lanham: Rowman & Littlefield.

Rosset, Peter Michael, Braulio Machin Sosa, Adilén María Roque Jaime, and Dana Rocío Ávila Lozano. 2011. The Campesino-to-Campesino agroecology movement of ANAP in Cuba: Social process methodology in the construction of sustainable peasant agriculture and food sovereignty. *The Journal of Peasant Studies* 38 (1): 161–191.

Safrin, Sabrina. 2004. Hyperownership in a time of biotechnological promise: The international conflict to control the building blocks of life. *American Journal of International Law* 98: 641–685.

Schroeder, Doris. 2009. Justice and Benefit Sharing. In *Indigenous peoples, consent and benefit sharing: Lessons from the San-Hoodia case*, ed. Rachel Wynberg, Doris Schroeder, and Roger Chennells, 11–26. Dordrecht: Springer.

Schroeder, Doris, and Balakrishna Pisupati. 2010. *Ethics, justice and the convention on biological diversity*. Nairobi: United Nations Environmental Program.

Schroeder, Doris, and Thomas Pogge. 2009. Justice and the convention on biological diversity. *Ethics & International Affairs* 23 (3): 267–280.

Sen, Amartya. 2009. *The idea of justice*. Cambridge, MA: Harvard University Press.

Shaver, Lea. 2010. The right to science and culture. *Wisconsin Law Review* 121: 121–184.

Shrader-Frechette, Kristin. 2002. *Environmental justice: Creating equality, reclaiming democracy*. Oxford: Oxford University Press.

Simon, Herbert A. 2001. UBI and the Flat Tax. In *What's wrong with a free lunch?*, ed. Philippe Van Parijs, Joshua Cohen, and Joel Rogers, 34–38. Boston: Beacon Press.

Sklair, Leslie. 2016. The transnational capitalist class, social movements, and alternatives to capitalist globalization. *International Critical Thought* 6 (3): 329–341. https://doi.org/10.1080/21598282.2016.1197997.

Soria-López, Manuel, and Israel Fuentes-Páramo. 2016. The identification of biopiracy in patents. *World Patent Information* 47: 67–74.

Standing, Guy. 2016. *The corruption of capitalism: Why rentiers thrive and work does not pay*. London: Biteback.

Stephan, Paula E. 2012. *How economics shapes science*. Cambridge, MA: Harvard University Press.

Sterckx, Sigrid. 2006. The moral justifiability of patents. *Ethical Perspectives* 13 (2): 249–265.

Thomas, Gigesh, and Johan De Tavernier. 2017. Farmer-suicide in India: Debating the role of biotechnology. *Life sciences, society and policy* 13 (1): 8.

Timmermann, Cristian. 2014. An assessment of prominent proposals to amend intellectual property regimes using a human rights framework. *La Propiedad Inmaterial* 18: 221–253.

Timmermann, Cristian. 2015. Pesticides and the patent bargain. *Journal of Agricultural and Environmental Ethics* 28 (1): 1–19. https://doi.org/10.1007/s10806-014-9515-x.

Timmermann, Cristian. 2017. Harvesting the uncollected fruits of other people's intellectual labour. *Acta bioethica* 23 (2): 259–269. https://doi.org/10.4067/s1726-569x2017000200259.

Timmermann, Cristian. 2018. Food security as a global public good. In *Routledge handbook of food as a commons*, edited by José Luis Vivero-Pol, Tomasso Ferrando, Olivier De Schutter, and Ugo Mattei, 85–99. Oxon & New York: Routledge.

Timmermann, Cristian, and Georges F. Félix. 2015. Agroecology as a vehicle for contributive justice. *Agriculture and Human Values* 32 (3): 523–538.

Timmermann, Cristian, and Zoë Robaey. 2016. Agrobiodiversity under different property regimes. *Journal of Agricultural and Environmental Ethics* 29 (2): 285–303. https://doi.org/10.1007/s10806-016-9602-2.

Tittonell, Pablo. 2013. *Farming systems ecology: Towards ecological intensification of world agriculture*. Wageningen: Wageningen Universiteit.

Tittonell, Pablo, Laurens Klerkx, Frederic Baudron, Georges F. Félix, Andrea Ruggia, Dirk van Apeldoorn, Santiago Dogliotti, Paul Mapfumo, and Walter AH. Rossing. 2016. Ecological intensification: Local innovation to address global challenges. *Sustainable Agriculture Reviews* 19: 1–34.

Vail, John. 2010. Decommodification and egalitarian political economy. *Politics & Society* 38 (3): 310–346.

Vivero-Pol, José Luis, Tomasso Ferrando, Olivier De Schutter, and Ugo Mattei (eds.). 2018. *Routledge handbook of food as a commons*. Oxon & New York: Routledge.

Vossenkuhl, Wilhelm. 2006. *Die Möglichkeit des Guten*. München: C.H. Beck.

Waldron, Jeremy. 1992. Superseding historic injustice. *Ethics* 103 (1): 4–28.

Waldron, Jeremy. 2003. Indigeneity? First peoples and last occupancy. *New Zealand Journal of Public and International Law* 1: 55–82.

Walton, Andrew. 2014. Do moral duties arise from global trade? *Moral Philosophy and Politics* 1 (2): 249–268.

Wenar, Leif. 2015. *Blood oil: Tyrants, violence, and the rules that run the world*. Oxford: Oxford University Press.

Wenner, Danielle M. 2015. The social value of knowledge and international clinical research. *Developing World Bioethics* 15 (2): 76–84. https://doi.org/10.1111/dewb.12037.

Wield, David, Joanna Chataway, and Maurice Bolo. 2010. Issues in the political economy of agricultural biotechnology. *Journal of Agrarian Change* 10 (3): 342–366.

Wilson, James. 2007. GM crops: Patently wrong? *Journal of Agricultural and Environmental Ethics* 20 (3): 261–283.

Chapter 5
Distributing Research Attention in Global Agriculture

Abstract Nowadays market incentives have the strongest influence in the distribution of research attention. In a world of extreme inequality in purchasing power and research capabilities, this outcome has disastrous consequences for social justice. The strong incentives produced by markets divert research attention away from the urgent needs of the poor to some of the most banal cravings of the rich. One way to understand distributive justice is as demanding that fair shares of research attention be allocated in proportion to the urgency of global agricultural problems. Yet obliging people to do a particular type of work amounts to a form of conscription. The aim of this chapter is to deal with the advantages and disadvantages of understanding research attention as a distributive good.

Keywords Allocation of scientific efforts · Distributive justice · Innovation gap · Underserved areas

5.1 On the Highly Unequal Distribution of Research Attention in Agriculture

In relation to agricultural innovation, there are different ways in which we can apply principles of distributive justice to identify potential injustices. We could, for instance, understand distributive justice in a very narrow sense, and interpret this notion of justice as demanding a fair distribution of the objects of innovation, for example, by handing out improved seed varieties, agricultural machinery and tools on the basis of normative considerations. Yet in agriculture, in contrast with other research fields, this interpretation will have only limited results, as it clashes against a major viability problem: innovation needs vary with local environmental and social conditions, which limits their successful adoption in other social and natural environments (Mazoyer and Roudart 2006; Leach et al. 2012). This constraint invites us to look at another good that also falls under the scope of distributive justice: research attention. A fair distribution of research attention can provide innovative solutions for the geographic areas that currently need the highest amount of research attention: regions that are vulnerable to climate change and whose food production capacity

C. Timmermann, *Social Justice and Agricultural Innovation*, The International Library of Environmental, Agricultural and Food Ethics 31,
https://doi.org/10.1007/978-3-030-56193-2_5

is currently underdeveloped (Cline 2007). It could also concentrate research on the types of solution that are better suited to dealing with the problems of the poor, particularly inventions that can be cheaply acquired or replicated using locally available spare elements (De Winter and Kosolosky 2014; Vanloqueren and Baret 2009; Gupta 2010). This could address the so-called "availability problem": the sufficient provision of innovative solutions that are especially suited to overcome the problems of those in need.

Another, partially overlapping, understanding of distributive justice is one that is commonly used in the environmental justice literature (Figueroa and Mills 2001; Shrader-Frechette 2002), which, applied to our case, would call for a fair distribution of the burdens and benefits of agricultural innovation. In terms of benefits, this approach would also call for more research aimed at helping to overcome the problems that primarily affect the global poor. In terms of distributing the burdens of agricultural research, this would demand a fairer allocation of research, development, diffusion and conservation costs, and providing the innovations needed to reduce the unfair geographic distribution of those plantations that are not used for direct human consumption (e.g. animal feed and fibres) and compete with food production for valuable resources, such as land, water and labour.

Inspired by these two approaches, the central goal of this chapter is to examine whether research attention can be considered a distributive good, and if so, to find out whether justice requires a fairer distribution of research attention, especially in relation to making available innovations that address the agricultural problems of the most food-vulnerable people in the world.

There are a number of factors that have made it urgent to discuss the problem of the fair distribution of research attention. Nowadays, in a world dominated by intellectual property rights, markets set the highest financial rewards for inventions that are needed to develop the products sought by the richest consumers. This ultimately has the effect that the market ends up concentrating most of the overall research attention in the private sector, and increasingly also in academia and public institutions, as pressure to seek funding from industry and to develop public–private partnerships increases (Lacey 2012; Stephan 2012; Sterckx 2011). In a world of extreme inequality, the fact that market pressures reflect purchasing capacity and not the number of people in need—nor the urgency of their needs—creates major conflicts for the realization of social justice. We must recall that we live in times where financial inequality is reaching the highest levels in history (Piketty 2014). A small fraction of the world's population with huge purchasing capacity will attract nearly all research attention (Hollis and Pogge 2008; Love and Hubbard 2007; Hassoun 2015). The strong incentives that markets produce divert research attention away from the urgent needs of the poor to even the most banal cravings of the rich (Timmermann and van den Belt 2012; Reiss and Kitcher 2009). Issues that primarily affect people in the tropical regions—the area where poverty and hunger is most endemic—do not receive nearly enough research attention. In global health, this problem is known as "the 10/90 gap": the situation where 10% of health research resources are spent on the problems of developing countries, which is the region where 90% of all preventable deaths occur (Drugs for Neglected Diseases Working Group 2001; Flory and Kitcher 2004). A similar situation can be seen in agricultural research, where much of private research concentrates on ornamental varieties and commercially profitable crops, many of which are used as animal feed or biofuel, or are destined for the textile industry (De Schutter 2009). In practice, this means that scientists spend more resources on the aesthetic characteristics of ornamental varieties than on vital crops like cassava or millet, which feed people in the tropics. This highly unequal distribution of research attention has the consequence that major regions of the world that suffer hunger do not benefit enough from scientific advancement to improve their situation, something that we may recall is a human right (Chapman 2009; Timmermann 2014).[1] There are strong research needs to improve tropical agriculture in terms of increasing yields, recovering soil fertility, improving resilience to extreme weather events, reducing tedious and health-deteriorating work, social and environmental sustainability, matching food production to dietary needs and cultural values, and most importantly, ensuring adequate innovation adoption by both rich and poor farmers (McIntyre et al. 2009; Baulcombe et al. 2009).

To help support a fairer alignment of research attention towards the needs of the poor, we must develop strong arguments based on theories of social justice (Buchanan et al. 2011). The most straightforward ethical reasoning that one could use to argue for a fairer distribution of research attention is utilitarianism. The suffering caused by hunger and malnutrition is felt by a much larger number of people, and much more intensely, than is required to counterbalance any inconveniences that researchers and those profiting from their work may suffer, thus morally justifying a redirection of research attention towards the needs of the global poor. Such a realignment, however, clashes with a key objection from political liberalism. Obliging people to do a particular type of work amounts to a form of conscription. This is a particularly troublesome outcome, as this obligation will only affect a specific segment of society: one which already makes a considerable contribution to society by enlarging the future pool of knowledge available to humankind.

In this chapter, I examine the advantages and disadvantages of understanding the uneven allocation of research attention as an issue of distributive justice. I start by providing a general overview of the problems involved in understanding research attention as a distributive good, as well as its potential. I then highlight those considerations that support a utilitarian argument for shifting research attention towards the needs of the poor, and identify some of the additional issues that a utilitarian calculation needs to consider. After that, I show that distributing research attention will indeed substantially affect scientific and civic liberties. In order to support a fairer distribution of research attention, despite its negative effect on scientific liberties, I offer three additional lines of supporting argument. First, I argue for an obligation to contribute back to society in return for the use of public resources, based

[1] See the Universal Declaration of Human Rights (1948), art. 27.1: "Everyone has the right freely to participate in the cultural life of the community, to enjoy the arts and to share in scientific advancement and its benefits"; and the International Covenant of Economic, Social and Cultural Rights (1966), art. 15.1 (b): [the right of everyone] "To enjoy the benefits of scientific progress and its applications."

on principles of reciprocity. Second, I discuss how research funds could be made conditional on contributing to States' commitments towards securing human rights. Third, I highlight the importance of diversifying lines of research and maintaining trust in science by appealing to arguments based on the social value of science.

5.2 Distributive Injustice in the Allocation of Research Attention: Preliminary Considerations

The aim of distributive justice is to seek a morally acceptable distribution of benefits and burdens in society. In a narrower sense, it is understood as dealing with the fair distribution of existing resources on which people have legitimate claims (Schroeder and Pisupati 2010). The extent to which we can understand a distributive arrangement as just depends on various factors and moral perspectives. We need to start by evaluating on what grounds people have claims on agricultural innovation, according to their own needs and circumstances.

Egalitarians have been involved in a long-standing debate about what measure (or currency) should be used to assess equality: opportunities, welfare, resources, or capabilities (Anderson 1999; Dieterlen 2015). Fortunately, for present purposes we not need engage in such a theory-laden examination. The main objective of and justification for agricultural innovation is to feed the world. Here I adhere to human rights discourse, and understand the right to adequate food as requiring access to a sufficient, continuous and adequate supply of food. Sufficient and continuous food is tied to nutritional needs for living a flourishing life, and to meeting the requirements of special circumstances and life stages such as pregnancy, breast-feeding, childhood, and heavily demanding work. Adequate food takes also into consideration reasonable non-nutritional preferences, particularly those tied to cultural and religious traditions (UN Committee on Economic Social and Cultural Rights 2006). Nowadays we also need to consider preferences and interests based on secular ethics, such as concerns about animal welfare and wider ecological issues, particularly climate change and biodiversity loss (Gremmen et al. 2019). In terms of justice, this would demand that research attention related to agricultural innovation be distributed to such an extent that the right to adequate food is sufficiently met. To be clear, this right is not met if researchers focus on meeting calorific need without paying attention to reasonable cultural and religious preferences, environmental concerns and specific dietary needs. The distribution of any surplus resources would need to be justified using a different normative framework. Here we must remind ourselves that the world already produces enough food to feed every inhabitant (Tittonell et al. 2016).

As a general distinction, we can distinguish criteria that aim to provide a minimum to fulfil everyone's need and arguments that condemn the excessive use of resources. In the first case, we need a set of principles that would defend a distribution of research attention that guarantees a sufficient supply of food. Sufficientarianism is a branch of distributive justice that we can explore for this line of argument, as its aim is to secure a distribution of socially and individually important goods that allows people to live a flourishing life (Shields 2012). Alternatively, and suitably for agriculture, we could condemn outcomes where some people have too much—in terms of research attention, land, and water resources—as is done under limitarianism (Robeyns 2017). Due to resource constraints, dietary preferences that consume a massive amount of resources, for example by including large daily meat portions, do not allow others to secure their basic food needs. In terms of research attention, we may observe an excessive concentration on satisfying aesthetic ideals, and the perfection of highly specialized industries in order to offer, for example, more sophisticated varieties of cheese and wines at the cost of the needs of subsistence crops.

When justifying claims on global research attention, there are two general considerations that we need to address on behalf of those who are sceptical of obligations that go beyond our national borders. The two main objections to distributing goods beyond national borders is that there is no sufficient associative relationship, and that one country is not harming the other. In agriculture, however, we can observe both associative relationships and harm beyond borders.

There are several factors that cast doubt on the claim that we lack associative ties with people beyond our national borders (cf. Koller 2013; Tan 2006). To start, in our case, world agriculture has benefitted from the identification of edible varieties from around the world, and from a massive exchange of crop germplasm and domesticated species (Mazoyer and Roudart 2006). Moreover, food security is a global public good that has a number of positive externalities from which people who have not contributed cannot be excluded, ranging from moral benefits to public health gains, and from the expansion of markets to political stability (Timmermann 2018b). Global cooperative justice, in the sense of contributing one's fair share to the establishment of global public goods, could justify some distributive measures to avoid letting certain people or nations free-ride (Van Parijs 2012). This would establish some types of associative obligation for past, present and future common work. As mentioned in the first two chapters, it is practically impossible to assess how much each innovator has contributed to an invention: something which is even more difficult when genetic resources are involved that have been improved over millennia. Any attempt to quantify how much each ancestor contributed to the existence of our common pool of agricultural knowledge and resources would lead to a dead end. We would also need to qualify each contribution and its relative importance by a widely accepted metric: an impossible task. It is safe to say that our current agricultural common heritage was only made possible by collecting knowledge and resources from all over the world, and that it is now impossible to dismantle this commons with reference to prior contributions (Frison and Coolsaet 2018).

In the case of not harming others, we already have a major game changer with the problem of climate change (Gilson and Kenehan 2018). Most of the countries who are now in a position to help others have made a much greater historical contribution to climate change and have thus harmed other countries, particularly those in the tropics who are most affected (Cline 2007; Giddens 2009; Meyer and Roser 2010). In addition, there are pollutants from agriculture and nitrogen excesses that cross borders (McIntyre et al. 2009).

Moreover, views to the effect that we should only compensate for the negative outcomes of bad luck, and not for those of bad choices, are consequently difficult to defend for agriculture. During our first two decades of life we can hardly be accountable for the choices we make, since we do not have a political voice in most countries. In later life, we should not underestimate the baggage of past bad decisions and the problem of path dependency. Changing our agricultural system is hardly a choice that can be fulfilled by one generation alone. Even in our later years of life, many of the problems we face arise because of decisions that we did not make and could not influence. Here we should keep in mind the direct consequences of leaving people empty-handed. If we conclude that it is in any way justifiable to leave people without the benefits of scientific advancement, this may lead to hunger. This would mean that people's bad choices, or even their ancestors' bad choices, may lead to hunger and yet it would be morally acceptable to refrain from helping people in such cases, due to the choices made. Here we need to be clear that, as long as we abide by human rights, inflicting hunger is not an acceptable form of punishment (cf. Anderson 1999).

Another argument, coming from a utilitarian perspective, is that a shift in research attention under current science expenditure would not amount to a major loss of quality of life. Governments could easily double science expenditures without worsening their citizens' quality of life, leaving half of that expenditure for business as usual and the other half for the global poor. Before exploring this line of argument further, let us discuss some of the feasibility issues related to treating research attention as a distributive good.

Moving from a theoretical problem to the allocation of a resource for securing a fundamental need, we need to take the following factors into consideration:

(1) One central justification for agricultural innovation is to ensure that people have enough access to adequate food.
(2) People can only eat a certain amount of food, which makes endless accumulation for personal consumption pointless.
(3) Moreover, food spoils, which obliges us to secure a continuous supply of food.
(4) It is impossible to trace exactly how much each region has contributed to improving agriculture with its knowledge or genetic resources.

Let us now move on, keeping these considerations and distinctions in mind.

5.3 On Understanding Research Attention as a Distributive Good

Generally speaking, research attention is a vital resource due to its role in providing solutions to life-threatening problems, through technological and social innovations. Yet, as we have seen, in a world of extreme inequality in purchasing power, and with substantial differences between the basic needs of poorer populations and the wants of richer populations, market incentives are unhelpful for distributing research attention fairly. Justifying a poverty-sensitive distribution of research attention is a complex task, and for this we need to examine if and to what extent research attention can really be considered a distributive good. To do so, I first examine what this supply depends on and, second, address practical problems for its distribution: the comparability, transferability and divisibility of research attention.

Concerning the total supply of research attention—the good in question to be distributed—it is important to keep in mind that we are not dealing with a fixed good. The total amount of research time available depends on public choices concerning national and international politics, political interference by interest groups, and the decisions and values that people have as individuals. These factors change the amount of resources available, and the economic and social incentives to do scientific work and perform well. As these factors determine the quantity and quality of resources available for distribution, they are not morally neutral, for individual communities can thereby, under certain circumstances, rightfully prioritize the needs of their own people before assisting others. This is the case especially when we are comparing similarly wealthy countries. Therefore we should take a closer look at these factors, as they are highly relevant for the realization of distributive justice.

As mentioned before, much scholarship on distributive justice has been concentrated on the distribution of resources whose quantity is relatively fixed, such as exhaustible natural resources. In contrast, the case of distributing research time is particularly difficult to assess, as we must consider three major factors: (1) the total number of spare resources a country may have to spend on science, (2) the policies that determine the amount of research time available, and (3) social appreciation of scientific work. Differences in policy and the public appreciation of science has led to a situation where one country may end up spending up to ten times as much on research as other countries at a similar development stage, as revealed by World Bank indicators.

For present purposes, let us understand "research attention" as the total number of work hours spent by researchers. The total amount of research attention that can potentially be made available depends upon multiple factors. As an upper limit, the hours of research attention available depend upon the number of people who are capable of doing research and the hours of research that they are able to do. Researchers have to be sustained using resources that are of limited supply. Depending on the direction in which research heads, the resources needed for science can be enormous, as seen in big science projects, such as satellite observation systems and whole genome sequencing. Moreover, these resources need to be produced and maintained, reducing the potential number of hours available for research. Also, for research attention to be available, researchers first need to be trained: something that takes a considerable amount of time, both for those in training and for those who provide the training, as well as further resources—both again, inputs of limited supply. In theory, by considering such factors, we could calculate the numbers of people that can be sustainably hired as researchers. Of course, our global society is very far from reaching any such upper limit on the number of researchers. Currently no country spends more than 5% of its GDP on research and development, while the average research expenditure in OECD countries was 2.49% in 2016.[2]

The factor that most constrains the amount of research attention available is, however, neither potential capacity nor resource availability. There are a number of obstacles in our political institutions which keep resources from flowing into scientific projects. Voters regularly support candidates who claim to want to improve the innovation capacity of their country, but fail to secure the resources needed for achieving this goal, or even cut research spending once in office. In some cases, there is simply no candidate with an effective science agenda who has any real chance of being elected. Natural or social disasters may also force politicians to drastically change government budgets for several years. Moreover, politicians are rarely held accountable for not fulfilling their campaign promises, or even for doing the opposite of what they promised. Here we should also not forget that many countries do not allow people to participate in the design of public policies. Moreover, increasing total research output is only one of many goals that a country may have. Public support for science also depends on the level of scientific literacy, and on how far the public perceives science as a source of solutions for urgent problems that is worth supporting. This leaves research budgets in a precarious situation in countries that do not see science as a priority or even as a reasonable investment opportunity, as can be observed in most parts of the world.

Here we also need to remember that scientific capacity is a good that takes a considerable amount of time to develop. It takes a lot of time to build the mixture of financial, social and moral incentives needed to encourage people to engage in science. The social standing of science in society may either encourage people to become scientists, or to continue to do scientific work, or dissuade them from doing so. Factors as diverse as lack of credibility, social stigma, an inability to combine research with family and social life, and poor salaries and working conditions discourage people from committing to long-term scientific work. It takes considerable effort to build trust among citizens so that they become willing to make the substantial investment needed to involve themselves in scientific careers and support scientific institutions. A willingness to make investments in one's education is easily eroded when scientific qualifications are not properly recognized, and researchers are rarely hired on merit (Iliescu et al. 2018). Similar factors affect the inclusiveness of scientific institutions. Do they have sufficient and continuously available funding to allow the integration of people from low-income families? Is there a professional code that demands the inclusion of people from different backgrounds? Have institutions established strong codes of conduct to prevent sexual harassment and discrimination? Are there scientific role models who members of under-represented groups can identify with? All these factors affect the number and the diversity of people who are willing to involve themselves in scientific efforts.

Choices not only affect the *quantity* of research attention available, but also the *quality* of research attention. This poses a major problem for distributive justice. To be able to distribute a good, we need to be able to make comparisons. In other words,

[2]See research and development expenditure (% of GDP), at https://data.worldbank.org/.

if research attention *A* does not produce the same output as research attention *B*, we need at least to be able to quantify this difference in order to fairly allocate the overall resource "research attention" among those parties who have a legitimate claim to it.

Here again we can observe multiple factors that influence the quality of research attention, something which will ultimately affect the justification for its distribution, i.e. to develop solutions for those in need.

As a matter of justice, and in order to secure good long-term performance and diversity, researchers should have certain labour rights and a significant amount of freedom of inquiry. This requires financial resources to secure decent wages, and also institutions, laws and procedures to counter corruption and scientific malpractice, and guarantee the protection of labour rights. Enough funding is also needed to pay those involved in research for their work, including graduate students and support staff, so as to prevent overworking and exploitative work arrangements, and to allow mobility and scientific exchanges. Maintaining an efficient, hospitable and safe research infrastructure, where people can (and would want to) work for long hours during all seasons, also costs money. The long-term sustainability of an innovation system also depends on the possibility of balancing work, leisure time and family life: something that will require us to disincentivize working for excessive hours and limit mobility expectations.

Here it is also important to channel sufficient resources to improve the efficiency of research institutions and reduce malpractice. This requires the establishment of policies to fight discrimination and promote fair hiring schemes, and strong mechanisms to enforce these policies. Nepotism in both public and private research institutes has an enormous negative effect on the quality of research, as it leads to a massive waste of resources and opportunities, and to epistemic poverty, resulting from a failure to diversify staff members (Wray 2007). The partial implementation of such policies, alongside group efforts to improve research capacities, can explain the enormous differences that can be observed, at both national and international levels, between resources put into research (e.g. wages, incentives and infrastructure) and research output (e.g. indexed peer-reviewed publications per researcher, citations). Fighting discrimination and improving the quality and efficiency of research institutes will reduce the extremely high difference in performance seen nowadays, allowing greater comparability between the research attention provided by different institutions.

The quantity and quality of research time available not only depends on regional or national policies, but in many instances is also strongly affected by decisions made abroad. Scholars of decolonization have provided a detailed analysis of how colonial policies either prohibited or discouraged the development of manufacturing industry in areas outside the main territory that were intended to be used only by extractive industries. Colonies were often prohibited from trading with each other, and education was mostly provided out of taxes (if at all) up to a level that was insufficient for developing skills for a specialized industry or the sciences (Galeano 1971/2008; Rodney 1972). Early colonial science had little interest in exploring any elements that would not serve the imperial industry: something that has massively delayed the exploration of local priorities (Schiebinger 2004). Many countries with a colonial past have still not succeeded in establishing a knowledge exchange among themselves, especially at the intercontinental level between Africa, Latin America and Asia, as can be observed in data analysis on scientific collaboration (Mazloumian et al. 2013).

In addition, there is the problem of foreign interventions. For example, the numerous interventions of the United States in Latin America had major negative consequences for the latter's scientific landscape. They helped to create a hostile environment that led to the large-scale murder, torture and expulsion of Latin American leftist and oppositional intellectuals. Military regimes were generally unsupportive of scientific programmes and closed whole faculties and research programmes when they had any links to leftist groups or agendas that could somehow be interpreted as socialist, such as social medicine or even nutritional studies (Galeano et al. 2011). Many military regimes had strong ties to right-wing religious groups: distorting research agendas, discriminating against women and indigenous people in science, hindering the participation even of relatively liberal and socially oriented religious groups, and often even impeding research on the effectiveness of liberal policies, indigenous intellectual contributions and social programmes (Dussel 1983). We can observe similar consequences around the world when fundamentalist or military regimes receive illegitimate support from foreign governments. We must take into consideration such international politics and historical injustices when assessing issues of global distributive justice.

Now for the problem of *transferability*. How should we understand the transferability requirement for distributive goods? Here we need to make some distinctions. Basic research is generally transferable, as it seeks to understand general scientific matters. Research tools are also transferable, in the sense that they can be used in different scientific experiments. In a strict sense, research results are generally transferable, in the sense that they have to be replicable. However, their usefulness may vary. For example, a method to improve the growth of apples is of little use in areas that do not meet the environmental conditions such trees require. For research results to be transferable, people need to develop the skills needed to convey these results in such a way that they can be understood by others. The recipients of such research also need training to understand the transmitted messages.

As far as the transfer of research attention from one area to another goes, people need to develop the skills needed to be receptive to the new research objectives and their social and environmental target areas, and to learn about previous scientific work and identify global priorities. There are substantial epistemic barriers that may impede researchers from developing technological solutions that are socially acceptable, suitable to the local environment, and apt to work in resource-poor settings (Anderson 2015). A shift in research attention requires a substantial amount of study, to acquire familiarity with the new subject matter and the ability to think from a wider range of perspectives, in order to analyse the local realities of target areas. This involves substantial costs to retrain researchers, in both time and money. This makes research attention a transferable good with substantial transition costs. As Herbert Simon observed, in countries in the Global North it is common to spend three times as much on transferring knowledge (or educating) as on creating knowledge (financing research and development), and this does not even take into account that the people who are being educated rarely get paid (Simon 1999).

Of major concern for agriculture is that a substantial amount of research has to be conducted on site, in order to test how an invention performs under the environmental and social conditions it is designed for (Wolf 2008). Due to existing global inequalities, this will involve a substantial amount of travel and relocation, both for the researchers from the Global North who go to underserved areas, and for the researchers from many countries of the Global South who go to places where they can receive an internationally competitive higher education (cf. Shaheed 2012). Without this mobility, research is strongly constrained by epistemic barriers.

On what the problem of *divisibility* of research attention is about, we find differences in opinion. While some people in biotechnology find the modularity of research efforts fascinating, as seen for example in synthetic biology, where scientists work to provide building blocks (biobricks) for custom assembling (O'Malley et al. 2008), researchers using ecosystems approaches have condemned the study of independent modules for leading to a negative form of reductionism, and to oversimplified models that do not take into account the overall effects to the human and ecological environment and their interaction (Max-Neef 2005; Thompson 2017; Anderson et al. 2019). A fair distribution of research attention only leads to good social and scientific results if the segmentation of research still delivers valid results. For example, it would be more important to conduct a single experiment with enough sample numbers for statistical significance, than to conduct several different experiments with tiny sample numbers that only leave open questions and do not allow us to draw conclusions. Some experiments require a far larger amount of resources due to their nature.

Those familiar with the discourse on rationing healthcare might be tempted to suggest that we should focus on the impact of innovation upon society when we distribute research attention, much as public health interventions are regularly measured with estimations based on quality-adjusted life years (QALY) or disability-adjusted life years (DALY) (cf. Hollis and Pogge 2008; Hassoun and Herlitz 2019). There are, however, severe problems with taking this approach in relation to agricultural innovation. As mentioned in the introductory chapter, agricultural innovations need to be ideally adapted to local socio-ecological circumstances, and ecological prudence entreats us to desist from using innovations (particularly regarding improved plant varieties) that are too extensively used. These factors call for a diversity of innovations, or even for incremental innovations. Rewarding innovations with a large impact would incentivize homogeneity, which would be contrary to many socio-ecological needs.

Considering the substantial limitations that exist in relation to the comparability, transferability and divisibility of research attention, we must concede that here we are dealing with a good that is highly unsuited to be distributed following precise guidelines. We can only have rough approximations that rely on imprecise and often inadequate proxies, such as the amount of money spent on research, the number of people working in the area, or the number of publications. All these metrics are problematic for assessing research. Yet we still can appeal to moral obligations for scientists and institutions to roughly assess how they allocate their attention, and to consider realigning their agendas when they are too detached from the needs of the poor.

5.4 Utilitarian Arguments for Redirecting Research Attention

According to a standard utilitarian argument, justice generally requires us to be willing to sacrifice something of smaller value if doing so will promote the greater good (Singer 1972). In our world as it currently stands, this type of utilitarian reasoning requires scientists and their funders to set aside private interests when deciding on research areas, as they ought to concentrate on solving the problems of the needy. The number of people who suffer hunger and malnutrition exceeds by far the number of people that professionally engage in research. The intensity of the suffering caused by hunger also exceeds by far any inconveniences involved in having to adjust research agendas to meet urgent social needs (cf. Elver 2016). As simple as this argument sounds, it makes a number of assumptions that need to be carefully assessed.

First, it assumes that people working in research areas that do not directly solve the problems of the needy could easily shift their research attention to improving the well-being of the worst off. This, however, is not the case. Here we are required to take the costs of retraining into consideration, which may lead to substantive gaps in research productivity. The argument also assumes that researchers will be sufficiently efficient doing other types of research, including research that may only benefit people from outside one's community. This assumes that there are no substantive knowledge gaps about site-specific circumstances. Here researchers need to maintain not only their research abilities, but also their motivation. As long as people are somewhat altruistic, this change could even increase their motivation, as research would improve the situation of the worst off: a demand found in many conceptions of justice and worldviews.

Second, the demand to shift research attention towards the needs of the poor comes at the price of a loss of liberty. This scientific freedom must not be valued too highly, so as to not amount to a disincentive to engage in research altogether. Traditionally, there has been strong support for defending scientific liberty as a crucial element of well-functioning innovation systems. Countless examples in the history of science show that strong government intervention in science has resulted in serious welfare issues. By this understanding, the scientific system performs at its best when scientists are allowed freedom of inquiry. For example, as seen in Chap. 3, one of the Mertonian norms requires scientists to be disinterested—in the sense that their only guiding principle is the quest for truth—and to ignore any other personal conviction (Merton 1973). Following such a principle would improve the quality of scientific work, and consequently a well-functioning scientific system will ultimately offer the best performance in terms of contributing to social welfare. Yet here we should remind ourselves that scientists from public institutes and universities nowadays already waive a considerable amount of freedom in order to secure research funding from industry (Macfarlane and Cheng 2008). As discussed in the first three chapters, the market is rapidly displacing scientific curiosity as the main factor that guides research orientation in most areas. Doing years of research, only to end up helping to adjust the colour of an ornamental plant variety, or any other such banal craving, may not be as strong an incentive as working to develop tools to fight world hunger. Cases where scientists are mainly driven by curiosity are becoming increasingly rare.

Third, it is often assumed that if one redistributes resources, one thereby improves the situation of some by reducing the amount of resources that others will have. While in a very immediate sense this is true, it can be seen that this does not have to be the case when we take indirect factors into consideration. Contributions to food security are also contributions to a global public good (Timmermann 2018b). Taking money away from research in ornamental varieties will have a minimal social impact, but reallocating this money to research into tropical food crops can produce multiple added advantages in the long run. Among the indirect benefits of reducing hunger and malnutrition through improved varieties, we can list the reduction in the amount of care work needed, due to people having fewer sick relatives; as adequate nutrition makes people less vulnerable to disease and more able to work and study, and leaves a higher share of the population able to provide for and help others. The use of better adapted varieties can also have positive environmental effects, by promoting better crop coverage, thereby retaining soil fertility and reducing erosion, and making the population less dependent on food imports that have to be transported from remote locations. When assessing redistributive policies from a utilitarian perspective, we should not forget the multiplying effect of poverty reduction measures. In some cases, redistribution can in the long run improve the situation of all, by shifting people from a state of dire need and dependency to a state of well-being and productivity. From an economic perspective, we might also condemn the situation where we cannot engage in mutually beneficial commercial and intellectual exchanges with the large portion of the world that is not integrated into markets or communication networks (Homann 2007).

In addition, the current allocation of research funding and industry development also has detrimental effects on the better-off population. Conventional agriculture has massive negative effects on our environment, thus affecting our mental and physical health. Intellectual property regimes have allowed major corporations to gain massive control over food systems through trademarks, protected industry standards and massive demand for homogeneous produce. Moreover, the massive levels of obesity and malnutrition currently seen do not reflect well-working systems (Otero et al. 2018).

In philosophical discourse, this line of argument is continuously accused of being overly demanding. At least at the beginning, it requires a substantial mass of people committing to such a position, to allow people enough space to follow more their banal cravings and other personal interests. Fortunately, the more people commit to advancing social welfare, the less demanding it becomes to address dire needs (Gomberg 2016; Singer 2009).

5.5 Obligations to Serve Underserved Areas: Reciprocity-Based Arguments

Another way to defend a fairer allocation of research attention is by borrowing from the discussion on brain drain and duties to work in underserved areas. Instead of interpreting brain drain solely as consisting of researchers physically migrating from poorer areas to richer areas, we can also qualify as brain drain the phenomenon of researchers scientifically moving from a research area that focusses on the needs of the poor to serving richer markets.

The most common objection to a duty to serve underserved (research) areas consists of characterizing this duty as a form of conscription. In contemporary political philosophy, John Rawls, for example, reserved conscription just for cases where a similar threat to personal liberty is at stake, such as an invasion by an enemy army (Rawls 1999). In practice, however, governmental policies, particularly in European countries with a strong social security system, were until recently much more open to making civil service mandatory, often reserving this as an option for those unwilling to do military service. We can see a similar compulsory element when funding is made conditional—even for goods protected by human rights articles, such as the human right to education. Many scholarship programmes are conditional: for example, in medicine, many student funding opportunities come with future obligations to serve in public hospitals; often even requiring recipients to move to distant rural areas. Many countries have policies that oblige people who use public funds to study abroad to later serve, during a specified time, back in their home country. Yet in these cases, conscription is in one way or another justified by appealing to reciprocity. In the case of civil service, each generation assumes a share of the social burden of caring for dependent people, partly in the hope of being taken care of themselves when the time comes. When accessing public funding, funds are made conditional on paying off the social debt by serving the taxpayers. Even when education is seen as a human right, it is not provided unconditionally. While recognizing the intrinsic value of developing literacy, democracies rely on the skillsets that are developed using public funds to facilitate informed voting, and to empower citizens to protect democratic institutions. Obligations such as practicing one's profession and remaining in the country that financed one's studies can also be seen as linked to principles of reciprocity, as it is crucial that those people who were able to get an education thanks to the existence of certain social structures then work to maintain those structures, in order to secure the conditions needed to grant the next generation their right to education (cf. Yuksekdag 2018).

Justifying an obligation to focus our research attention on the problems of the distant needy—i.e. the needy who live outside our national or regional borders—on principles of reciprocity is somewhat more difficult. One option is to continue to claim that agricultural researchers build on and use genetic resources that are owned by the public at large as a common heritage of humankind. Such use could be made conditional, on the basis of a hypothetical contract between society and the scientific community. As the now highly industrialized countries have benefited from millennia of work done by people who lived in areas that are now in need (Thrupp 2000; Engels et al. 2011), we could stipulate an obligation to give back to the people who now live in those areas. Similarly, one could argue for a duty to engage in the research needs of the poor by offering such research services as a way to compensate smallholders for agrobiodiversity conservation. As much of agrobiodiversity is conserved and enhanced by landraces in the Global South, thanks to the efforts of small farmers (De Jonge 2009; Timmermann and Robaey 2016), the scientific community could use a considerable amount of scientific resources to compensate this major effort, by providing innovations specially tailored to their circumstances (cf. Bragdon 2016).

A general problem with reciprocity-based arguments is that individuals tend to overvalue their own contributions, while at the same time undervaluing other people's input. This creates a false perception of seeing a fair system as unfair, even though objectively one should not see it this way. The global scientific system needs to be perceived as a fair cooperative endeavour, in order to make sure that people contribute to this system due to their own internal motivations, and not solely to avoid penalties. To reduce the effect just mentioned, massive educational campaigns are needed, especially to educate researchers on the enormous amount of previous work they are building on, and the work done by smallholders to conserve biodiversity and traditional knowledge. This will require us to dismantle the myths that give too much credit to individual innovators without acknowledging the roles of other intellectual contributors and social structures.

5.6 Obligations to Serve Underserved Areas: States Parties' Commitments to International Human Rights Treaties

Another way to argue for an obligation to shift research attention towards solving the problems of the needy, is to claim that research funds are given by the public for the specific task of improving social welfare. There are statements in various constitutions and charters that indicate or imply that science is a means to foster social welfare. The most influential case is article 11.2 of the International Covenant on Economic, Social and Cultural Rights (1966), in which we can read:

> The States Parties to the present Covenant, recognizing the fundamental right of everyone to be free from hunger, shall take, individually and through international co-operation, the measures, including specific programmes, which are needed:

(a) To improve methods of production, conservation and distribution of food by making full use of technical and scientific knowledge, by disseminating knowledge of the principles of nutrition and by developing or reforming agrarian systems in such a way as to achieve the most efficient development and utilization of natural resources.

(…)

Here it can be understood that improving social welfare should include helping people in need all over the world (De Schutter 2011). Doing science using public funds could be made conditional on abiding by this international commitment. States can partly fulfil their international commitments to help the poor by funding science meant to develop solutions for the needs of the poor. Here we need to keep in mind that such commitments are not a thing of past, only affirmed in reaction to the catastrophic humanitarian consequences of World War II, but in fact have been repeatedly restated. In the 1990s they can be found in the World Food Summit and the Millennium Development Goals (Lappé et al. 2013), and in the mid 2010s in the Sustainable Development Goals (Bórquez Polloni and Lopicich Catalán 2017).

There are some problems with this approach as well. The first problem is a political one. In global political arenas, the right to benefit from scientific advancement is one of the most neglected human rights (Chapman 2009; Marks 2011; Shaver 2010). Most international organizations and NGOs have instead focussed on the human right to food, as this is a basic need, and is something that people need to secure before they can enjoy other protected human rights (Shue 1996). This brings in complications. While the link between food and hunger is self-evident, the link between science and hunger is less direct. Having access to the latest scientific advances is not the only option—under current social and environmental conditions—for making food more accessible today. When focussing on immediate needs, it would be more prudent to start by reducing the amount of food that ends up wasted (Gjerris and Gaiani 2013), or is underused by feeding animals for human consumption (Singer 1975/2009; Oteros-Rozas et al. 2019). The situation of many farms in areas with hunger is so precarious that currently available technologies, such as irrigation systems and improved seeds, would suffice for major increments in harvests to alleviate short-term needs (Tittonell et al. 2016). Many agricultural innovations that are freely available have not been adopted by farmers, even when they have been developed especially for their local conditions (Röling 2009). Agroecological extension services could share know-how based on traditional knowledge and thereby also significantly improve harvest yields, as well as making farms more resilient to climate change and extreme weather events (Altieri and Koohafkan 2008). As long as existing methods of reducing food waste and improving yields are not exhausted, shifting research attention towards the needs of the poor becomes a mere means—and not even a very efficient one—for fighting today's hunger. As long as existing technologies are underused, we only have a weak argument for limiting scientific liberties. Curtailing scientific liberty might be only justifiable when this is strictly necessary to achieve a wider social goal, such as eradicating hunger. Another problem is that hunger can also be prevented by direct food aid, instead of agricultural development aid. When immediate action is given absolute preference over long-term sustainability, food aid will continuously be preferred over agricultural development (for criticism, see Thompson 2010). Shifting research agendas would only be therefore justifiable when this is linked to long-term development programmes backed by principles of environmental sustainability and food sovereignty.

In a strict sense, this argument would, however, put substantial pressure on social scientists and scholars in the humanities to work on social innovations to improve technology diffusion and uptake—by developing better teaching methods, improving theories of social responsibility, and providing stronger arguments for charity, volunteering and working for the common good.

Nonetheless, the fact that we can do so much to fight hunger at this moment without redirecting research is due to the prior injustice of our current food systems, where inequalities and deprivation are extreme. Nature continuously reacts to agricultural systems, and this will make innovation at all stages necessary again, and it is wise to already align our research agendas to prepare for future changes and secure sufficient harvest yields in a socially and environmentally sustainable manner. As soon as we can observe a better uptake of existing innovations, we will have a much stronger case for realigning technological innovation agendas towards the need of the poor.

Lastly, we need to note that this argument is relatively weak when applied to the private sector, particularly while people remain free to work in areas that do not have a strong social mandate, such as the liquor and cosmetic industries.

5.7 Commitments to Well-Ordered Science

There are major concerns about the distortion of science by corporate interests. A well-ordered science should have adequate assessment of technology and direct research to efficiently and sustainably solve social problems (Reiss and Kitcher 2009). This includes the evaluation of the efficacy of both preventive and reparative measures, independently of their profit opportunities (Intemann and de Melo-Martín 2014). It requires inclusive deliberation with the general public, to evaluate needs and their urgency, different solutions and approaches, and risks and benefits. To allow for public deliberations, a number of scientists will have to act as a bridge between scientists and the wider public, so that research agendas become aligned to public interests and needs, and trust in science is established (Hisano 2005). Such a dialogue would ideally allow research agendas to be shifted to address wider social interests, with appropriate consideration given to global challenges and the needs of the world's most vulnerable.

A problem of this approach is that research will then be aligned to the demands corresponding to people's prevailing ideas of justice. Fortunately, people from developed countries tend to overestimate the amount of funds that their countries spend on foreign development (Singer 2004), which indicates that, despite the current lack of sympathy for the global poor, public deliberation would in principle be open to spending much more on helping those in need than what is currently spent. However, we also need to be aware that science needs stability, and populist agendas often cut science spending drastically. Until now, candidates who have made eradicating global poverty a priority in their political agendas have fared poorly in elections.

One could also argue under this approach that there is scientific value in maintaining diversity by leaving at least some funding for reasonable alternative research lines. Then if mainstream approaches were to fail, the scientific community would have access to already developed research lines, and would not have to start again from a too early stage. Alternative approaches and knowledge acquisition methods can also help to shine a different light on mainstream research lines by identifying new problems and opportunities, as well as reconfiguring the discussion when apparent dead ends are reached.

This line of argument may only offer limited support for diversifying research lines and providing sufficient public solutions for the needy, so as to maintain public cooperation.

5.8 Some Additional Insights from Environmental Justice

As mentioned before, the environmental justice literature understands distributive justice as the fair sharing of benefits and burdens. We saw the complexities involved in sharing the benefits of agricultural innovation, in particular regarding productivity. How about sharing the burdens of agricultural innovation? What does this imply? Does this imply a fairer division of the costs of research and development? For example, should countries commit to a minimum in research spending in proportion to their GDP?

Moreover, we may ask ourselves: should we understand "burdens" indirectly? In this sense, our food systems unfairly burden certain areas of the world with the massive production of goods such as soy and palm oil, which contribute little to local food security and have an enormous negative impact on environmental sustainability. Researchers would have to take care that innovation does not end up overburdening some regions with cultivars not meant for food production. Such decisions should be empirically well-informed. If, for example, we want to assess the social and ecological desirability of a crop such as oil palms, we need to give appropriate weight to the interests of local populations (Castellanos-Navarrete and Jansen 2018), without forgetting to balance those interests with eventual costs to be paid by others, such as the crop's contribution to obesity by cheapening processed foods, and the large-scale loss of invaluable ecosystems.

5.9 Concluding Remarks

At this stage, we can see that the current highly inequitable distribution of research attention is a situation that, intuitively, can quickly be identified as a major global injustice, yet unfortunately it requires solving major normative and practical challenges. Although, in principle, research attention can be considered a distributive good, it will take time and much effort to make a transition that meets the demands of distributive justice. Even when applying simple utilitarian reasoning to this problem, we should not be blind to the multiple factors that may act as disincentives, and could even reduce overall welfare by discouraging people to become innovators in welfare-improving areas.

The case for dedicating research attention to reciprocating historical efforts, ecosystem services and conservation work has strong normative potential, as we currently do not compensate such efforts in any satisfactory way. We might also justify research spending on the development of innovative solutions for those in need on the basis that states would thereby comply with some of their human rights commitments, although this approach is argumentatively weaker for failing to sufficiently address private research. Another promising approach is to employ ideas of well-ordered science and its public mission to improve welfare and retain public support.

Moreover, when moving human capital to underserved areas, special care has to be taken to avoid mistakes like those that were made with international development aid. Here too, we must learn from historical lessons. Providing scientific services should involve local capacities, in order to help them to improve their efficiency, expand communication networks and contribute to self-sustenance. Scientific aid should not have the effect of eroding traditional ecological knowledge systems and procurement methods. It is worrisome how quickly traditional knowledge is lost before being properly assessed (Kelbessa 2015; Robinson 2008). Bringing in new experts from abroad should also not cause the local population to unjustifiably believe that experts from abroad are, merely due to their origin, more trustworthy than local experts (cf. Dübgen 2012).

Lastly, we have also duties, as a matter of contributive justice, to work towards solving the problems of those with the most urgent needs (Timmermann 2018a). We will discuss this notion of justice in the next chapter.

References

Altieri, Miguel A., and Parviz Koohafkan. 2008. *Enduring farms: Climate change, smallholders and traditional farming communities*. Penang: Third World Network.

Anderson, Elizabeth S. 1999. What is the point of equality? *Ethics* 109 (2): 287–337.

Anderson, Elizabeth. 2015. Feminist epistemology and philosophy of science. In *The Stanford encyclopedia of philosophy*, ed. by Edward N. Zalta. Standford: Standford University.

Anderson, Colin Ray, Janneke Bruil, Michael Jahi Chappell, Csilla Kiss, and Michel Patrick Pimbert. 2019. From transition to domains of transformation: Getting to sustainable and just food systems through agroecology. *Sustainability* 11 (19): 5272.

Baulcombe, D., I. Crute, B. Davies, J. Dunwell, M. Gale, J. Jones, J. Pretty, W. Sutherland, C. Toulmin, and N. Green. 2009. *Reaping the benefits: Science and the sustainable intensification of global agriculture*. London, UK: The Royal Society.

Bórquez Polloni, Blanca, and Boris Lopicich Catalán. 2017. La dimensión bioética de los Objetivos de Desarrollo Sostenible (ODS). *Revista de Bioética y Derecho* 41: 121–139.

Bragdon, Susan H. 2016. Reinvigorating the public sector: The case of food security, small-scale farmers, trade and intellectual property rules. *Development* 59 (3–4): 280–291.

Buchanan, Allen, Tony Cole, and Robert O. Keohane. 2011. Justice in the diffusion of innovation. *Journal of Political Philosophy* 19 (3): 306–332. https://doi.org/10.1111/j.1467-9760.2009.00348.x.

Castellanos-Navarrete, Antonio, and Kees Jansen. 2018. Is oil palm expansion a challenge to agroecology? Smallholders practising industrial farming in Mexico. *Journal of Agrarian Change* 18 (1): 132–155. https://doi.org/10.1111/joac.12195.

Chapman, Audrey. 2009. Towards an understanding of the right to enjoy the benefits of scientific progress and its applications. *Journal of Human Rights* 8 (1): 1–36.

Cline, William R. 2007. *Global warming and agriculture: Impact estimates by country*. Washington, DC: Center for Global Development.

De Jonge, Bram. 2009. Plants, genes and justice: An inquiry into fair and equitable benefit-sharing. Ph.D. diss., Applied Philosophy Group, Wageningen University.

De Schutter, Olivier. 2009. Seed policies and the right to food: Enhancing agrobiodiversity and encouraging innovation (Report presented to the UN General Assembly, 64th session, UN doc. A/64/170).

De Schutter, Olivier. 2011. The right of everyone to enjoy the benefits from scientific progress and the right to food: From conflict to complementarity. *Human Rights Quarterly* 33: 304–350.

De Winter, Jan, and Laszlo Kosolosky. 2014. Health, food, and science: An ethical assessment of research agendas. *Logique et Analyse* 57 (228): 701–726.

Dieterlen, Paulette. 2015. *Justicia distributiva y salud*. Ciudad de México: Fondo de Cultura Económica.

Drugs for Neglected Diseases Working Group. 2001. Fatal imbalance: The crisis in research and development for drugs for neglected diseases. Geneva: Médecins Sans Frontières.

Dübgen, Franziska. 2012. Africa humiliated? Misrecognition in development aid. *Res Publica* 18: 65–77.

Dussel, Enrique D. 1983. *Praxis latinoamericana y filosofía de la liberación*. Bogotá: Nueva América.

Elver, Hilal. 2016. The challenges and developments of the right to food in the 21st century: Reflections of the United Nations special rapporteur on the right to food. *UCLA J. Int'l L. Foreign Aff.* 20: 1–40.

Engels, Johannes, Hannes Dempewolf, and Victoria Henson-Apollonio. 2011. Ethical considerations in agro-biodiversity research, collecting, and use. *Journal of Agricultural and Environmental Ethics* 24 (2): 107–126. https://doi.org/10.1007/s10806-010-9251-9.

Figueroa, Robert, and Claudia Mills. 2001. Environmental justice. In *A companion to environmental philosophy*, ed. by Dale Jamieson, 426–438. Malden & Oxford: Blackwell.

Flory, James H., and Philip Kitcher. 2004. Global health and scientific research agenda. *Philosophy & Public Affairs* 32 (1): 36–65.

Frison, Christine, and Brendan Coolsaet. 2018. Routledge handbook of food as a commons. In *Routledge handbook of food as a commons*, ed. by José Luis Vivero-Pol, Tomasso Ferrando, Olivier De Schutter, and Ugo Mattei, 218–230. Oxon & New York: Routledge.

Galeano, Diego, Lucía Trotta, and Hugo Spinelli. 2011. Juan César García y el movimiento latinoamericano de medicina social: Notas sobre una trayectoria de vida. *Salud colectiva* 7: 285–315.

Galeano, Eduardo. 1971/2008. *Las venas abiertas de América Latina*. Madrid: Siglo XXI.

Giddens, Anthony. 2009. *Politics of climate change*. Cambridge: Polity.

Gilson, Erinn Cunniff, and Sarah Kenehan (eds.). 2018. *Food, environment, and climate change: Justice at the intersections*. Lanham: Rowman & Littlefield International.

Gjerris, Mickey, and Silvia Gaiani. 2013. Household food waste in Nordic countries: Estimations and ethical implications. *Etikk i praksis-Nordic Journal of Applied Ethics* 7 (1): 6–23.

Gomberg, Paul. 2016. Why distributive justice is impossible but contributive justice would work. *Science & Society* 80 (1): 31–55. https://doi.org/10.1521/siso.2016.80.1.31.

Gremmen, Bart, Vincent Blok, and Bernice Bovenkerk. 2019. Responsible innovation for life: Five challenges agriculture offers for responsible innovation in agriculture and food, and the necessity of an ethics of innovation. *Journal of Agricultural and Environmental Ethics* 32 (5–6): 673–679.

Gupta, Anil K. 2010. Grassroots green innovations for inclusive, sustainable development. In *The innovation for development report 2009–2010, strengthening innovation for the prosperity of the nations*, ed. by Augusto Lopez-Claros, 137–146. Houndmills & New York: Palgrave Macmillan.

Hassoun, Nicole. 2015. The global health impact index: Promoting global health. *PLoS ONE* 10 (12): e0141374.

Hassoun, Nicole, and Anders Herlitz. 2019. Distributing global health resources. Contemporary issues in political philosophy. *Philosophy Compass* 14 (11): e12632.

Hisano, Shuji. 2005. A critical observation on the mainstream discourse of biotechnology for the poor. *Tailoring Biotechnologies* 1 (2): 81–106.

Hollis, Aidan, and Thomas W. Pogge. 2008. *The health impact fund: Making new medicines accessible for all.* Oslo & New Haven: Incentives for global health.

Homann, Karl. 2007. Globalisation from a business ethics point of view. In *Globalisation and Business Ethics*, ed. Karl Homann, Peter Koslowski, and Christoph Luetge, 3–9. Aldershot: Ashgate.

Iliescu, Adrian-Paul, Ileana Dascălu, Thierry Ngosso, and Naomi van Steenbergen. 2018. Intergenerational justice in the context of developing countries. In *Towards the ethics of a green future*, ed. Marcus Düwell, Gerhard Bos, and Naomi van Steenbergen, 110–130. London: Routledge.

Intemann, Kristen, and Inmaculada de Melo-Martín. 2014. Addressing problems in profit-driven research: How can feminist conceptions of objectivity help? *European Journal for Philosophy of Science* 4 (2): 135–151.

Kelbessa, Workineh. 2015. African environmental ethics, indigenous knowledge, and environmental challenges. *Environmental Ethics* 37 (4): 387–410.

Koller, Peter. 2013. Social and global justice. In *Spheres of global justice: Volume 1 global challenges to liberal democracy. Political participation, minorities and migrations*, ed. by Jean-Christophe Merle, 433–443. Dordrecht: Springer.

Lacey, Hugh. 2012. Reflections on science and technoscience. *Scientiae studia* 10 (SPE): 103–128.

Lappé, Frances Moore, Jennifer Clapp, Molly Anderson, Robin Broad, Ellen Messer, Thomas Pogge, and Timothy Wise. 2013. How we count hunger matters. *Ethics & International Affairs* 27 (03): 251–259.

Leach, Melissa, Johan Rockström, Paul Raskin, Ian Christopher Scoones, Andrew C. Stirling, Adrian Smith, John Thompson, Erik Millstone, Adrian Ely, and Elisa Arond. 2012. Transforming innovation for sustainability. *Ecology and Society* 17 (2): 11.

Love, James, and Tim Hubbard. 2007. The big idea: Prizes to stimulate R&D for new medicines. *Chicago-Kent Law Review* 82 (3): 1519–1554.

Macfarlane, Bruce, and Ming Cheng. 2008. Communism, universalism and disinterestedness: Re-examining contemporary support among academics for Merton's scientific norms. *Journal of Academic Ethics* 6 (1): 67–78. https://doi.org/10.1007/s10805-008-9055-y.

Marks, Stephen P. 2011. The neglected human right to benefit from scientific progress: implications for human development. Human Development and Capabilities Association 2011 International Conference, Den Haag, 6–8 September 2011.

Max-Neef, Manfred A. 2005. Foundations of transdisciplinarity. *Ecological economics* 53 (1): 5–16.

Mazloumian, Amin, Dirk Helbing, Sergi Lozano, Robert P. Light, and Katy Börner. 2013. Global multi-level analysis of the 'Scientific Food Web'. *Science Reports* 3 (1167). https://doi.org/10.1038/srep01167.

Mazoyer, Marcel, and Laurence Roudart. 2006. *A history of world agriculture: From the neolithic age to the current crisis.* New York: Monthy Review Press.

McIntyre, Beverly D., Hans R. Herren, Judi Wakhungu, and Robert T. Watson. 2009. *International assessment of agricultural knowledge, science and technology for development (IAASTD):*

Synthesis report with executive summary: A synthesis of the global and sub-global IAASTD reports. Washington, DC: Island Press.
Merton, Robert K. 1973. The normative structure of science. In *The sociology of science*, ed. by Norman W. Storer, 275–77. Chicago & London: The University of Chicago Press.
Meyer, Lukas H., and Dominic Roser. 2010. Climate justice and historical emissions. *Critical review of international social and political philosophy* 13 (1): 229–253.
O'Malley, Maureen A., Alexander Powell, Jonathan F. Davies, and Jane Calvert. 2008. Knowledge-making distinctions in synthetic biology. *BioEssays* 30 (1): 57–65.
Otero, Gerardo, Efe Can Gürcan, Gabriela Pechlaner, and Giselle Liberman. 2018. Food security, obesity, and inequality: Measuring the risk of exposure to the neoliberal diet. *Journal of agrarian change* 18 (3): 536–554.
Oteros-Rozas, Elisa, Adriana Ruiz-Almeida, Mateo Aguado, José A. González, and Marta G. Rivera-Ferre. 2019. A social–ecological analysis of the global agrifood system. *Proceedings of the National Academy of Sciences*. https://doi.org/10.1073/pnas.1912710116.
Piketty, Thomas. 2014. *El capital en el siglo XXI*. Mexico, DF: Fondo de Cultura Económica.
Rawls, John. 1999. *A theory of justice*. Rev. Cambridge, MA: Belknap Press of Harvard University Press.
Reiss, Julian, and Philip Kitcher. 2009. Biomedical research, neglected diseases, and well-ordered science. *THEORIA. Revista de Teoría, Historia y Fundamentos de la Ciencia* 24 (3): 263–282.
Robeyns, Ingrid. 2017. Having too much. In *Wealth: NOMOS LVI*, ed. Jack Knight and Melissa Schwartzberg, 1–44. New York: New York University Press.
Robinson, Daniel. 2008. Beyond 'protection': Promoting traditional knowledge systems in Thailand. In *Patenting Lives: Life Patents, Culture and Development*, ed. Johanna Gibson, 121–138. Aldershot: Ashgate Pub.
Rodney, Walter. 1972. *How Europe underdeveloped Africa*. London: Bogle-L'Ouverture Publications.
Röling, Niels. 2009. Pathways for impact: Scientists' different perspectives on agricultural innovation. *International journal of agricultural sustainability* 7 (2): 83–94.
Schiebinger, Londa. 2004. Feminist history of colonial science. *Hypatia* 19 (1): 233–254.
Schroeder, Doris, and Balakrishna Pisupati. 2010. *Ethics, justice and the convention on biological diversity*. Nairobi: United Nations Environmental Program.
Shaheed, Farida. 2012. Report of the special Rapporteur in the field of cultural rights. United Nations (A/HRC/20/26). Geneva: UN Human Rights Council.
Shaver, Lea. 2010. The right to science and culture. *Wisconsin Law Review* 121: 121–184.
Shields, Liam. 2012. The prospects for sufficientarianism. *Utilitas* 24 (1): 101–117.
Shrader-Frechette, Kristin. 2002. *Environmental justice: Creating equality, reclaiming democracy*. Oxford: Oxford University Press.
Shue, Henry. 1996. *Basic rights: Subsistance, affluence, and U.S. foreign policy*, 2nd ed. Princeton, N.J.: Princeton University Press.
Simon, Herbert A. 1999. The many shapes of knowledge. *Revue d'économie industrielle* 88 (2): 23–39.
Singer, Peter. 1972. Famine, affluence, and morality. *Philosophy & Public Affairs* 1 (3): 229–243.
Singer, Peter. 1975/2009. *Animal liberation*. New York: HarperCollins.
Singer, Peter. 2004. *One world: The ethics of globalization*, 2nd ed. New Haven & London: Yale University Press.
Singer, Peter. 2009. *The life you can save: How to do your part to end world poverty*. New York: Random House.
Stephan, Paula E. 2012. *How economics shapes science*. Cambridge, MA: Harvard University Press.
Sterckx, Sigrid. 2011. Patenting and licensing of university research: Promoting innovation or undermining academic values? *Science and engineering ethics* 17 (1): 45–64.
Tan, Kok-Chor. 2006. The boundary of justice and the justice of boundaries: Defending global egalitarianism. *Canadian Journal of Law & Jurisprudence* 19 (2): 319–344.

Thompson, Paul B. 2010. Food aid and the famine relief argument (brief return). *Journal of Agricultural and Environmental Ethics* 23 (3): 209–227.
Thompson, Paul B. 2017. *The spirit of the soil: Agriculture and environmental ethics*. New York: Routledge.
Thrupp, Lori Ann. 2000. Linking agricultural biodiversity and food security: The valuable role of agrobiodiversity for sustainable agriculture. *International affairs* 76 (2): 283–297.
Timmermann, Cristian. 2014. Sharing in or benefiting from scientific advancement? *Science and Engineering Ethics* 20 (1): 111–133.
Timmermann, Cristian. 2018a. Contributive justice: An exploration of a wider provision of meaningful work. *Social Justice Research* 31 (1): 85–111. doi: https://doi.org/10.1007/s11211-017-0[illegible]
Timmermann, Cristian. 2018b. Food security as a global public good. In *Routledge handbook of food as a commons*, ed. by José Luis Vivero-Pol, Tomasso Ferrando, Olivier De Schutter, and Ugo Mattei, 85–99. Oxon & New York: Routledge.
Timmermann, Cristian, and Zoë Robaey. 2016. Agrobiodiversity under different property regimes. *Journal of Agricultural and Environmental Ethics* 29 (2): 285–303. https://doi.org/10.1007/s10806-016-9602-2.
Timmermann, Cristian, and Henk van den Belt. 2012. Global justice considerations for a proposed "Climate Impact Fund" . *Public Reason* 4 (1–2): 182–196.
Tittonell, Pablo, Laurens Klerkx, Frederic Baudron, Georges F. Félix, Andrea Ruggia, Dirk van Apeldoorn, Santiago Dogliotti, Paul Mapfumo, and Walter AH. Rossing. 2016. Ecological intensification: Local innovation to address global challenges. *Sustainable Agriculture Reviews* 19: 1–34.
UN Committee on Economic Social and Cultural Rights. 2006. General Comment No. 17: The right of everyone to benefit from the protection of the moral and material interests resulting from any scientific, literary or artistic production of Which He or She is the Author (Art. 15, Para. 1 (c) of the Covenant, E/C.12/GC/17). Geneva: United Nations Economic and Social Council.
Vanloqueren, Gaëtan, and Philippe V. Baret. 2009. How agricultural research systems shape a technological regime that develops genetic engineering but locks out agroecological innovations. *Research Policy* 38 (6): 971–983.
Van Parijs, Philippe. 2012. International distributive justice. In *A companion to contemporary political philosophy*, edited by Robert E. Goodin, Philip Pettit and Thomas W. Pogge, 638–652. Malden & Oxford: Blackwell.
Wolf, Steven A. 2008. Professionalization of agriculture and distributed innovation for multifunctional landscapes and territorial development. *Agriculture and Human Values* 25 (2): 203–207.
Wray, K. Bradley. 2007. Evaluating scientists: Examining the effects of sexism and nepotism. In *Value-free science: Ideal and illusions?*, ed. Harold Kincaid, John Dupré, and Alison Wylie, 87–106. Oxford: Oxford Univesity Press.
Yuksekdag, Yusuf. 2018. Health without care? Vulnerability, medical brain drain, and health worker responsibilities in underserved contexts. *Health Care Analysis* 26 (1): 17–32.

Chapter 6
Contributive Justice and the Importance of Inclusive Agricultural Innovation Systems

Abstract We can observe an increasing dissatisfaction about the lack of inclusivity in both the processes and governance of agricultural innovation. To argue for increased participation, I rely on the idea of contributive justice to identify the different requirements for establishing an inclusive research environment. Under these requirements we can list (i) opportunities to participate, (ii) opportunities to develop skills, (iii) opportunities to learn how to be productive, (iv) a duty to contribute towards social welfare, (v) allowing a fair competition of ideas, (vi) establishing a fairer distribution of meaningful and tedious tasks, and (vii) providing recognition on a non-discriminatory basis. This chapter examines how these demands apply to agricultural innovation.

Keywords Inclusiveness · Participation · Open innovation · Recognition justice · Participatory development · Self-determination

6.1 In Support of Participation: An Introduction

While the so-called accessibility and availability problems of innovation have received the most attention in the literature so far, the societal goal of increasing participation in science and technology development is usually treated as a secondary priority, to be left aside until more urgent needs such as access to food and healthcare are sufficiently covered. To address the fundamental human interests in participating in scientific projects and their governance (Shaver 2010; Chapman and Wyndham 2013; Donders 2011), particularly in relation to other interests that are commonly seen as more urgent, we may use the notion of contributive justice to build a normative framework that argues for wider inclusion and capacity-building in agricultural innovation (Timmermann and Félix 2015; Timmermann 2018).

The concept of contributive justice has received renewed interest in social justice research since the early 2000s, in order to argue against the hurdles created by racial discrimination, class advantage and poverty for obtaining meaningful work and developing skills (Gomberg 2007; Sayer 2009, 2012; Britz and Lipinski 2001).

C. Timmermann, *Social Justice and Agricultural Innovation*, The International Library of Environmental, Agricultural and Food Ethics 31,
https://doi.org/10.1007/978-3-030-56193-2_6

According to the interpretation here defended, we can state six demands of contributive justice that are particularly relevant for agricultural innovation: opportunities to participate as a peer, opportunities to develop skills, opportunities to learn how to be productive, a duty to contribute to social welfare, a fair competition of ideas, and doing one's share of tedious tasks (Timmermann 2018). To these we can add, due to historical discrimination against smallholders, the need to offer recognition on a non-discriminatory basis. These demands concern both the direction of research and how it is carried out.

The aim of this chapter is to show how the idea of contributive justice can be used to demand more inclusive innovation environments as a matter of social justice. This will be done by discussing whether we should neglect inclusive development and capacity-building while focussing on more urgent problems, such as preventing famine, and then examining how the seven aforementioned demands of contributive justice apply to agricultural innovation.

6.2 Do We Have to Choose Between Capacity-Building and Food Security?

To state the obvious, people cannot benefit from other goods protected by human rights without continuous access to sufficient food. Due to physiological limitations, hungry people cannot function at their full capacity. Furthermore, people who fear not having enough food for themselves and their loved ones usually dedicate their full attention to finding food, and this hinders them from enjoying a flourishing life in its full variety (Caparrós 2014; Ziegler 2011a). This physiological limitation has led to the well-known distinction between basic needs and other human rights. This division is standardly taken to imply that basic needs have to be secured before we focus on less urgent human rights that can be addressed at a later stage of social development. The reasoning behind this inference is that some goods are necessary to be able to benefit from other goods. To give a straightforward example, hunger strikes aside, one needs to eat before one can participate in long-term political activities.

Some strong objections have been raised against this reasoning. The first objection we can list is that raised by Shue (1996), according to which sufficient access to cover essential needs cannot be sustainably and continuously secured unless we simultaneously secure a right to participation. Basic needs such as food and water will always be in jeopardy, as long as people cannot secure a meaningful participation in civil society and democratic decision-making processes to defend their interests and rights. In the realm of agriculture, this would mean that we need, as a matter of justice, an inclusive system that allows people to participate in decision-making processes and scientific enterprises, and offers transparency in these processes. Such openness will also allow the public to identify, to a much greater extent, different forms of corruption and inefficient governance, such as a failure to hire the best people available in the field, the illicit use and improper exploitation of commonly held resources, and the bribing of public officials (Atenas et al. forthcoming). As Sen (1981) showed in his classic essay on famine, information and public participation have a key instrumental value for avoiding famine and large-scale hunger, and this is one reason why he identified freedom of the press as crucial in the fight to end hunger.

A second objection is raised by the people and organizations who advocate for food sovereignty. Here the right to adequate food is seen as tied to a right to produce food, while deciding sovereignly over production processes (Via Campesina 1996). Access and control over the means of production is demanded as a central human interest, not only for its instrumental value, which, in order to be meaningful, would require access to land, water and agricultural innovations (Nyéléni Forum for Food Sovereignty 2007; Rosset 2009). Food production and agricultural innovation have to be subject to decisions reached on the basis of open deliberations. Addressing these demands would require openness in innovation processes, as a right, in order to allow for self-determination and increased participation: something that should not be postponed until after basic needs are secured (Timmermann et al. 2018).

The third line of reasoning starts from a human rights perspective. On a broad understanding of human rights, as observed in the Universal Declaration of Human Rights (1948), participation in science qualifies as a human right. Article 27.1 reads: "Everyone has the right freely to participate in the cultural life of the community, to enjoy the arts and to share in scientific advancement and its benefits" (Timmermann 2014; Vayena and Tasioulas 2015). Here science is also seen as part of our common cultural heritage. Taking into account current global wealth and the avoidability of hunger (Lappé et al. 2013), we are not in the position of having to decide between fulfilling an urgent human right and securing other less urgent human rights. We easily have enough resources in the world to secure both human rights. By this reasoning, a failure to secure one human right (the right to adequate food) should not be used as a justification for failing to secure another human right (the right to share in scientific advancement).

Based on these three lines of arguments, there is a strong case that the right to food cannot and should not be pursued independently of initiatives seeking increased participation in innovation and its governance.

6.3 Contributive Justice in Agricultural Innovation

The line of reasoning I follow in this chapter builds on the idea of contributive justice. This form of social justice condemns insufficient opportunities to participate in agricultural innovation, not only because of the negative consequences of an exclusive innovation system, but because it demands opportunities to participate in science, as an issue of justice. As mentioned before, we can find such reasoning in the Universal Declaration of Human Rights (1948), in Article 27. According to this article, people should have a right to participate in cultural life, which includes being able to actively participate in scientific advancement (Chapman and Wyndham 2013;

Besson 2015; COMEST 2015). From this perspective, people have a strong interest not only in being mere spectators of the advancement of science, but also in being active participants (Timmermann 2014; Plomer 2012). Here it is worth mentioning that participating in science is not only satisfactory for a curious mind, but also allows people to develop tools that can significantly improve one's own and other people's well-being (Timmermann and Félix 2015). Inclusiveness in agricultural research is particularly important, since agriculture is the profession that involves the most people, including 2.5 billion smallholders (Holt-Giménez and Altieri 2013), and has the largest impact on the terrestrial environment, extending to over 70% of the Earth's land mass (Röling 2009). Moreover, as long as we want to live, we cannot avoid being consumers of agricultural produce. As a consequence, we cannot boycott agricultural products as a whole, and therefore are necessarily complicit in the social and environmental damages that food systems cause. It is a matter of social justice to empower people to transform these systems into fair enterprises.

To argue for more inclusive innovation systems, I will rely on a theory of justice that has regained scholarly interest in the last two decades: the idea of contributive justice. This concept involves a series of demands to increase the quality and quantity of participation. In earlier work (Timmermann 2018) I have identified six elements that can be considered as the central demands of this notion of justice:

(1) Opportunities to participate
(2) Opportunities to develop skills
(3) Opportunities to learn how to be productive
(4) A duty to contribute to social welfare
(5) Fair competition of ideas
(6) A fair distribution of tedious and meaningful tasks.

Due to a widespread failure to acknowledge certain types of innovations, and certain groups of people who innovate, I here include an addition demand:

(7) To offer recognition on a non-discriminatory basis.

In the following subsections, I apply each of these demands in order to highlight a series of social justice issues regarding agricultural innovation processes. These demands serve both to make sure that people are in a position to contribute to social welfare by making use of specially acquired and developed skills, and also to create an environment that is receptive to these contributions. A world where either people cannot contribute to social welfare, or their contributions are not taken seriously, is likely to be perceived as unjust.

6.3.1 Opportunities to Participate

As an initial demand of contributive justice, we can name an obligation to offer opportunities to participate. Generally speaking, this entails a social mandate to make innovation processes inclusive, by making them open and avoiding unnecessary complexities in order to facilitate wider participation. Participation is seen as a fundamental human interest, and scientific projects are not exempt from this demand of justice, as they are a central influence on society and are a source for recognition. Opportunities to participate should also include having a say in science governance. As noted by early scholars of technology assessment, it is fundamental that democratic decision-making, especially in a technological era like the one we live in, also embraces science governance (Olivé 2013). Participation in agricultural research is of special concern. As mentioned before, every human being needs food to live, so we are unable to avoid consuming food products. How this sector is organized affects us all, directly by the objects it produces and indirectly by its impact on the environment. We cannot avoid being accomplices in the wrongdoings that are triggered and indirectly caused by our food consumption. Facilitating participation in agricultural innovation is therefore a matter of social justice, as it is a fundamental human interest to avoid having to contribute to unjust institutions, and to be able to redress the institutions we rely on (cf. Young 2004).

There are multiple barriers to improving participation. Current levels of inequality make certain redistributive measures mandatory for improving participation. First of all, hunger and malnutrition are the most urgent problems for improving participation in innovative projects and their governance. Hunger is not only a problem for school children learning the basic knowledge needed to participate in democracies, but also in higher education institutions. Food insecurity—in terms of real hunger, inappropriate diets and worries about obtaining food—is a major impediment to attentively participating in research projects, for a large proportion of university students in both the Global South and the Global North (Ngumbi 2019). Second, as food sovereignty advocates list in their demands (Via Campesina 1996), infrastructure in rural areas needs to be significantly improved so that farmers can fully participate in civil society. The trend within conventional agriculture of moving innovation processes away from the field to specialized laboratories needs to be countered, either by reversing this trend, or by distributing experimental research sites across rural areas and improving the diffusion of communication technologies so as to allow remote participation. People, especially farmers, should also be able to access advanced training facilities in rural areas. Similar efforts also need to be made in agroecology to link creative minds with each other so that they can cooperate in problem solving (Tittonell 2019). Here the drastic reduction of costs related to communication technologies can be of great help in narrowing the digital divide, and widening access to training opportunities and scientific knowledge (Schoonmaker 2007). The building of rural digital learning centres can help bridge the digital divide by facilitating access to digital infrastructure (Gupta 2006). A failure to improve access to the Internet and improve digital literacy will greatly increase inequalities, as the cost of lacking continuous access to this pool of knowledge is a major deficit in a world where people extract enormous advantages from this pool of information and data, as well as the networking possibilities that these communication technologies offer (Johnson 2014; Moss 2002; Britz 2004).

Another obstacle to openness is created by proprietary science systems (Torrance and von Hippel 2015). To maintain patent eligibility, scientists have a strong interest in keeping innovation processes confidential until exclusive rights can be sought, in order to meet the novelty requirements. The lengthy period of exclusivity associated with intellectual property rights impedes a timely and free sharing of ideas. Extreme inequality makes it impossible—or at least very difficult—to acquire licenses from richer parties, giving a systematic advantage to richer countries. Even small license fees can be prohibitively expensive for the over 2.5 billion smallholders who hardly have any disposable income at all. Moreover, the path dependence created by proprietary science incentives favours innovations that are difficult to replicate and improve on an individual basis and can therefore be constrained more easily, such as hybrid seed varieties (Kloppenburg 2005).

Discrimination also limits participation. Discrimination has to be actively tackled, since participation, in order to be meaningful, requires that one can participate as a peer whenever one has the required training or relevant knowledge (Fraser 1998). Here we need to be conscious of the multiple effects of discrimination based on gender, race, religion and ethnic background. To maximize participation, it is not only important to overcome actual discrimination and biases, but also to allow people from underrepresented groups to regain the confidence to express themselves in a style that they are comfortable with and is appropriate to the message they need to deliver to specific audiences (Bayruns Garcia 2019), and this is crucial for hot topics such as hunger and the protection of the environment (Bovenkerk 2012). Discrimination consists of a double injustice: impeding or deteriorating the experience of participation, and also depriving the community of the benefits of cooperating in cognitively diverse environments.

The lack of diversity in scientific research teams remains a major social justice issue, as this excludes people and limits cognitive diversity: factors that reduce the quality of research outputs (Editorial 2018). The allocation of government funding to institutions that officially discriminate against certain religions (or irreligion) is still common practice, as seen, for example, in several German public universities who give the Vatican City State veto power over the appointment of academic chairs (Lotter 2007). Sometimes discrimination can manifest itself in less obvious ways. For example, when access and participation requires certified knowledge, the agencies who control and decide such certification can have a substantial amount of power (Ribot and Peluso 2003). This power can be used to maintain positions of privilege, and to oppress different forms of knowledge and knowledge acquisition, regardless of their cognitive and social value.

Innovation systems characteristically encourage others to build on previous successes. The pursuit of scientific projects is not an individual endeavour, but is rather strongly dependent on the opportunities that others have to participate in such projects (Mormina 2019). Science is a social enterprise (Jefferson 2006). In agriculture, this also requires that people in similar biological, climatic and social environments are able to participate. Individuals need to be able to share and exchange observations and samples with others who work in a similar environment. Moreover, a central motivation for scientific research, as with other cultural endeavours, is that others will continue to work on present findings in the future (cf. O'Neill 1993).

For participation to be fully inclusive, it has to embrace different forms of knowledge acquisition and contribution. Parallel knowledge systems increase possibilities of contributing to scientific advancement by opening participation opportunities in different physical locations, in both urban and rural areas, and through different knowledge-acquisition channels, from local observations to standardized scientific systems. Here the role of traditional knowledge systems becomes particularly important, due to the enormous number of users and contributors involved, as this allows a very large segment of the world's population to gather an enormous amount of knowledge through local observation and knowledge exchange systems (ICSU Study Group on Science and Traditional Knowledge 2002). To maximize the benefit of this resource pool, and to increase the degree and intensity of participation, a considerable amount of translation work is needed to document such findings in a common scientific language, so that findings can be analysed, replicated, further improved, and adapted to different environmental and social conditions. Under current levels of education, this would also require translating the most important information into local languages, so that it becomes accessible to people from different parts of the world, particularly those living in similar environments. Agroecologists have taken a major step here by working closely with smallholders: revising and optimising these techniques and transferring them to other social and ecological environments (Altieri and Koohafkan 2008). Other approaches aim to include citizens systematically in research, as seen in various citizen science initiatives (Eitzel et al. 2017; Wiggins and Wilbanks 2019; Fiske et al. 2019). Here citizens are valued for their input, while ideally also being given a say in the direction of research, as well as room to satisfy their intellectual curiosity (Timmermann 2019).

In scientific projects, attention needs to be paid to diversifying local scientific cooperation partners. For instance, when agricultural extensionists continuously work with the same farmers with whom they had success in the past, this may aggravate inequalities (Röling et al. 1976), and may also distort the extrapolation of results concerning user adoption and innovation. To avoid creating such contributive inequalities, it is important for people to diversify their research partners for the sake of inclusion, even when this comes at the cost of working in less effective teams and having to continuously develop new social bonds. Consciousness about justice needs to be strong, in order to counteract strong impulses to continue collaborating in the usual forms triggered by scientific curiosity, a sense of community, social ties, and a belief of contributing to the greater good.

6.3.2 Opportunities to Develop Skills

Being able to develop skills has both intrinsic and instrumental value for contributive justice. Learning a wide range of skills allows people to better help their communities, and being able to help one's community with the use of specially acquired skills is something that people value in itself. While it is possible to help one's community without the use of any specialized skills, the ability to do so with specialized skills often creates a situation where people can contribute to social well-being by doing meaningful work, instead of just doing necessary but tedious work. Therefore the ability to contribute to society with specially acquired skills combines meaningful work with people's desire to improve social welfare. Furthermore, the acquisition of skills makes one's contribution less replaceable, and sometimes even unique, which may lead to wider social and scientific recognition. In a more general sense, the development of skills is also crucial for changing the personal circumstances in which one finds oneself, something that may be considered a fundamental human need (Steinvorth 2009). This allows people to cooperate with others, in search of solutions to current social and environmental problems, to bring about wider transformations in society, or simply to develop and make use of places where like-minded people can find self-realization. Cooperation has the additional benefit of bringing people together, something that is often valued in itself. People may also value the development of skills for intrinsic reasons. This has been referred to as "the Aristotelian Principle", whereby people aim to continuously develop their skills for the sake of achieving virtuosity (Taylor 2004). Here the social recognition of the skill is secondary; more important is feeling content about gaining awareness of one's own mastery and working on the continuous improvement of one's skills.

In the case of agriculture, we must acknowledge that innovation is knowledge-intensive, and that we as a global society have developed a number of both tangible and intangible tools for facilitating and enhancing innovation processes. Here we have developed a series of methods and tools of inquiry and calculation—approximations for predictions, among others—that facilitate and expand scientific research capacities. Facilitating the development of skills is a major effort and consumes time. Moreover, the design of educational policies needs to take into account the interdependence of different knowledge domains. For a meaningful social and scientific engagement, people will have to acquire multiple types of knowledge, become scientifically and politically literate, learn to use information technologies and acquire knowledge of the local social and ecological environment. Increasingly, the complexity of scientific projects demands that people develop social and communication skills in order to do science and secure cooperation. It is important to underline the extensiveness and interconnectedness of the various skills or literacies that people will have to develop to allow meaningful participation. Nowadays, for instance, information literacy—in the sense of learning how to use, understand, exchange, share and retrieve information—is crucial for both scientific and political participation (Britz and Lor 2010), since informed consent requires access to information from trustworthy sources. Political literacy is of major importance in all development stages. People need to understand the different political channels that are available to push for political action, official procedures for issuing complaints and showing support, and the different possibilities and benefits of collective organization, so that they are empowered to change and defend democratic institutions and policies (de-Shalit 2004; Bornemann and Weiland 2019). Political literacy facilitates change, and allows people to defend political systems without resorting to violence and chaos. Most important for our topic is scientific literacy, as adequate participation in scientific advancement and technology assessment requires a good understanding of its processes, advantages and risks. The skills necessary for communication—either to solicit assistance, to exchange ideas or to share knowledge—are becoming increasingly important in our knowledge society (Lor and Britz 2005), even more so when we are talking about something as vital as improving food systems.

As pointed out in the highly influential work of Paulo Freire (1973), we must reconsider how skills are passed on. Being critical of the concept of extension, Freire highlights the importance of communication in education, and the importance of agronomists creating an environment in which farmers are able to acquire principles that they can further refine, as well as methods for improving knowledge acquisition and communication. In such educational settings, the agronomist is both an educator and a student, building a dialogue in which knowledge can be bidirectionally communicated and ultimately appropriated. Achieving such a receptive environment is a major challenge, as biases have to be overcome and farmers' knowledge and perspectives need to be revalorised, so that both agronomists and farmers are willing and able to learn from each other. Overcoming such difficulties is, however, crucial to educators improving themselves as educators, and farmers improving themselves as farmers; as well educators improving their success as farmers, and farmers improving their prospects as future educators.

To facilitate knowledge diffusion, famers' networks, such as the *campesino-a-campesino* movement, should be further promoted, to allow knowledge sharing and development among peers (Rosset et al. 2011). Successful farmers have great potential to teach their neighbours how to use innovations and solve problems, since they can show others on their farms the benefits of implementing such innovations, and in some cases might also be more trustworthy than outsiders (Šūmane et al. 2018). Farmers' networks have great potential for passing on knowledge from one farmer to another, but special care needs to be taken that skills are properly passed on—especially those involving more complex methods, ratios or associations—as knowledge is liable to quickly become distorted as it is passed along (Röling et al. 1976). It is important to create an environment where knowledge is not merely passed on linearly from progressive to less-progressive farmers, but where there is a reciprocal knowledge exchange, and that these knowledge exchange networks continuously aim to improve farms, revise innovations and critically test new ideas and methods.

When it comes to the transfer and development of skills, we should also not forget the social and socio-technical innovations we are currently witnessing. New forms of teaching and learning, exchanging ideas and organizing through the use of information technology have an enormous potential for making innovation inclusive. Agriculture, in particular knowledge-intensive sustainable agriculture, should make much better use of these new forms of interaction so as to increase the quality and intensity of participation and the number of voices from the Global South (cf. Coolsaet 2016). This also requires campaigns to be carried out to improve the social standing of these novel methods of learning and teaching, and give them the credit they deserve. Furthermore, governments should invest more in the exploration and development of these new educational methods, as they nowadays massively rely on the extracurricular activities of faculty members and the desperate labour circumstances of many early career researchers for the content and design of courses.

As far as intellectual property limitations are concerned, contributive justice would require open-access publications, or at least generous post-print archiving policies, so as to allow interested parties to develop their skills using the latest available scientific literature (Britz 2008). This requires a structural reform of scientific publishing and publication metric systems. Similarly, and of increased interest, freely available online courses and tutorials can be a great asset to assist the development of skills (Atenas and Havemann 2013). To develop and maintain skills, access to research tools is also needed, which in the case of agriculture also includes genetic resources, and this also requires improving the efficiency and reliability of postal delivery systems and customs. Exclusive rights should not hinder the training of researchers and farmers, and should not make lifelong learning harder.

While the existence of different streams of innovation allows for a wider spectrum of participants, this does not mean that facilitating participation in any one of these streams suffices. To secure genuine participation, people should have the liberty to choose to participate in any innovation stream, be it biotechnology or traditional knowledge systems, regardless of their background. This, of course, will require making training opportunities more widely available for all streams of innovation. The interest in participating is already there. As currently seen with the raise of agroecology, there is much interest from many agronomists from highly industrialized countries in studying traditional sustainable agriculture methods, while conversely, many agronomists from the Global South undergo specialized training in order to learn the tools and insights that biotechnology offers. Opportunities to participate need to be greatly expanded and made accessible to people from all backgrounds.

6.3.3 Opportunities to Learn How to Be Productive

Contributive justice also demands that work-saving techniques be taught. There is a considerable amount of research potential and need for developing work-saving tools that are especially suited to the social and environmental conditions of resource-poor regions, in particular for smallholders practicing sustainable farming methods. Here the development of tools that can be replicated using locally available materials needs to be prioritized, so that the poorest people around the globe can also benefit from innovation (Gupta 2010). In addition, labour-saving techniques and instruments for sustainable agriculture need to be developed to reduce the number and intensity of tedious tasks (Timmermann and Félix 2015). A failure to improve the productivity of smallholders may jeopardize their survival.

Research tools that could enhance productivity need to be developed and made accessible. This includes a wide array of technologies and machines necessary for agricultural research and the development of improved crop varieties. It also requires access to information, so that innovators can build on the state of the art instead of reinventing the wheel, so to speak. Access to information not only allows people to build on existing knowledge; it also provides greater insights and a wider overview. To employ an expression attributed to Isaac Newton, access to information allows us "to stand on the shoulders of giants". Moreover, by having up-to-date access to information, specifically scientific journal articles, researchers can identify potential research partners, facilitating collaboration and thus improving productivity. In today's world, scientists need to maintain a certain degree of productivity in many high-end research fields, for example concerning the quality and quantity of experiments, samples, publications or inventions, in order to participate as peers in scientific advancement and be eligible for competitive funding. This calls for measures to improve the productivity of researchers of the Global South by supporting infrastructure and sharing materials, know-how and information.

Another issue is that if we maintain parallel food systems, it becomes crucial for them to have a similar level of productivity. Nowadays we can observe huge differences in agricultural productivity. Half of the world's food production occurs in areas with yields below 3.1 t ha^{-1} $year^{-1}$, while 12.5% of the world's food production occurs in areas with yields above 6 t ha^{-1} $year^{-1}$, including areas with even 11 t ha^{-1} $year^{-1}$ (Tittonell et al. 2016). This disparity allows and even demands huge improvements, as it directly affects work quality and jeopardizes the survival of smallholders, both as people and as independent food producers. Progress in this dimension is taking place, however. Agricultural machinery is being developed for use on agroecological farms, through the use of new materials that are much lighter and designed to allow the removal of weeds without causing too much damage to the soil structure.

Concentrating our global attention on harvest yields when assessing productivity should not blind us to other relevant factors. While it is undeniable that farming systems need to be able to feed the people that work the land, our current ecological and social crisis obliges us to take a wider perspective than focussing on production alone (Gliessman 2007). When assessing a farm's performance, we should also take into account currently unpaid contributions to social welfare in terms of ecosystem services, such as urgently needed involvement in carbon sequestration and in situ biodiversity conservation (Thompson 2017). Measures of the success and social value of innovation should also take into consideration its broader effects on society and the environment.

6.3.4 A Duty to Contribute Towards Social Welfare

One highly criticized demand of contributive justice is the duty to contribute to social welfare (Gomberg 2016; Britz and Lipinski 2001). Taking this duty seriously in today's world would require aligning research interests with the needs of the poor. From this perspective, current involvement in ornamental varieties would be deemed unacceptable, given the levels of hunger prevalent in the world and the underperformance (in terms of yields) of tropical agriculture (cf. Tittonell et al. 2016). This would also oblige innovators and scientists to critically reassess the nature of their professions. The quest for truth would no longer be seen as the central goal of scientific research, but would instead have to go hand in hand with plans to improve social welfare.

Some interpreters of this type of duty embrace an obligation to contribute to social welfare that would even oblige people to develop the skills needed to help people in need (Gomberg 2007): something that has enormous potential to address the demands of global justice, but may involve substantial liberty-constraining costs for those who adhere to this principle (Parr 2015). In agriculture, this may ultimately also require moving to underserved areas, in order to carry out the needed field tests and acquire knowledge of local problems and opportunities. Considering the lengthy periods of time some field tests require, and the effects long absences have on social, family and romantic relationships, this is quite a demanding duty.

Particularly in relation to knowledge with a welfare-improving effect, contributive justice also demands that communities and nations share their knowledge (Lor and Britz 2005). No groups of people, be they corporations or indigenous communities, are exempt from this obligation. Sharing is emptied of meaning when the documented knowledge is only accessible to a limited group. This calls for information- and data-sharing policies that facilitate access to all interested parties with minimal costs. In this sense, scientific knowledge is claimed as part of the public domain, much like Merton's scientific ethos, as discussed in Chap. 3.

This duty faces major opposition from political liberals. We may recall that making contributions morally compulsory amounts to a form of conscription: something that, according to major political voices like Rawls (1999), is a form of social servitude that should only be used when proportional dangers to society are at stake, like the invasion of an enemy army. Despite the fact that hunger is a major threat to life and political stability (Holt-Giménez et al. 2010), famines are usually dealt with through food aid (which is often delivered by people conscripted into military service), but rarely combated with a systematic and immediate strategy that obliges the general population to perform certain tasks to improve food production systems. The standard response to hunger, as an immediate threat that would justify conscription under current political thought, is the delivery of food aid, and not the restructuring of food systems or the building of farms that would secure food after a few months. Conscripting people also to innovate to improve food systems would meet even higher opposition than obliging people to contribute to the (re-)establishing of food production capabilities with existing know-how.

As concerned citizens, we can still adhere to the duty to contribute to social welfare in a weaker and less demanding sense if this is interpreted as an ideal we should aspire to, but which nonetheless has a supererogatory character. This would invite people to critically reflect on their career choices in relation to their potential for improving social well-being. Another noteworthy issue, pointed out by Gomberg (2016), is that this duty becomes less demanding as more people follow suit. Nowadays, with the huge urgency and extensiveness of global problems and a richer global class that has become detached and sometimes even blind to the problems of the poor and the environment, the obligation to contribute to social welfare is massively overdemanding due to the magnitude of the problem, as even well-known utilitarian philosophers acknowledge (Singer 2009). Fortunately, this burden becomes smaller the more people embrace this principle of justice.

An open question is the extent of contributive justice's role in technology assessment. If we understand a duty to contribute to social welfare in a very strict sense, innovators would have to be especially concerned about the long-term and distant effects of their innovations. For instance, one accusation against conventional agriculture has been that it weakens the role and bargaining power of women in food production (Shiva 2009). In conventional agriculture it is predominately men who drive the tractors, apply agrochemicals and harvest the crops (Rosset et al. 2011). Once we realize that a stream of innovation disempowers women, we must shift our support to innovation streams that are more inclusive and redistribute power, and thereby contribute to overall social well-being and put more people in a position to help others. Judging by this example, depending on how strictly we interpret the concept of contributive justice, we will have different duties, which may also include a quite extensive examination of the potential long-term social and environmental effects of technology.

6.3.5 A Fair Competition of Ideas

Scholars from the Global South and providers of traditional ecological knowledge have long condemned discrimination within scientific communities (Kelbessa 2013)—and by people from the Global North in general (Rodney 1972)—as manifested by the testimonial injustice and epistemic injustice of which the former group are victims. Specifically in relation to agriculture, researchers in agroecology have accused supporters of conventional agriculture of unjustifiably discriminating against other scientific approaches, specifically ecological farming systems (Tittonell 2014). Such injustices involve the discrimination that certain types of informants may suffer for being part of a certain group or having certain characteristics, or due to how they have formulated their findings (Fricker 2007). In both cases, people are not given the same opportunities due to biases or acts of discrimination that others may direct towards them, or how they express themselves (Grasswick 2011). Scholars working on decolonization and feminist epistemology have done much work in beginning to unravel the extensiveness of such discriminatory interpretations, and uncovering the amount of people who never received due credit because of their origin, beliefs (or lack thereof), or for being women.

No branch of science should be unjustly favoured over any other. Similarly, certain knowledge-acquisition methods should not be absolutely preferred over others (Álvarez and Coolsaet 2020). This is unfair, not only to the people who acquire and depict knowledge in ways that differ from dominant approaches, but also to society, as alternative knowledge systems can identify important gaps in our current problem-solving approaches and provide solutions better suited to the people who will make use of them (Anderson 2015; Walajahi 2019). This may also include revoking the conceptual hegemony that is so strongly criticized by scholars of decolonization, as it obliges scholars from the Global South to frame and conceptualize their contributions and objections in terms that may not appropriately reflect their cultural background and thought (Escobar 1998).

An unfair competition of ideas is likely to demotivate people from contributing with their ideas and observations. Discrimination often has a deterring effect. Here we also have to take into account that even in open innovation systems, people do not share their innovations for solely altruistic reasons. The sharing of ideas may generate feedback and improve knowledge when others reciprocate by sharing secondary uses, identifying weak points and possibilities of improvement, thus refining innovation cycles (Gupta et al. 2016). Particularly positive or constructive feedback may encourage people to continue sharing, or encourage others to present and discuss their ideas as well.

Here some scholars have called for a fair competition of ideas (Ziegler 2011b; Sparrow and Goodin 2001) in order to stimulate the development of new and diverse ideas and thus break up the intellectual monopolies of well-established schools of thought. Traces of such views can also be found in Merton's norms of universalism and disinterestedness, as discussed in Chap. 3, which stress the need to make research environments inclusive for all according to their merit, and condemn the establishment of exclusive schools that prioritize certain groups of people and ideas (Merton 1942). A cognitively diverse environment would not only be more just but also intellectually richer (Kitcher 1990).

In agriculture, special attention needs to be paid to ensure that certain innovation efforts are not excluded from the competition of ideas—in the sense of fair evaluation by society and the scientific community—altogether (Šūmane et al. 2018). Innovations in areas where women have traditionally played a dominant role, such as the maintenance and development of landraces, are seldom recognized for their intellectual contribution and effort (Baquedano Jer and Larraín 2015; Shiva 2009). Since our innovation incentive systems do not reward these creative efforts, we need other ways to give adequate recognition to these knowledge systems, such as prizes and educational campaigns.

Establishing a globally fair competition of ideas has been very difficult, especially taking into account the extreme inequality between different research centres and systems. Publication outlets, conferences venues, media coverage, participation in trade fairs, and social media management have a huge influence on the diffusion of ideas, and to different degrees are all open to monetary influence, due to their commercial nature or dependency on subventions and third-party funding. Richer research institutes can influence all these venues and hire specialists to promote their ideas. Here again, vast amounts of financial capital can provide an unfair advantage and buy influence, illegitimately interfering in scientific practice and policy (cf. Robeyns 2017).

6.3.6 A Fair Distribution of Tedious and Meaningful Tasks

Another demand of contributive justice is for everyone to assume their fair share of tedious tasks, so as to avoid overburdening certain groups with such work, and to distribute meaningful work fairly (Gomberg 2007; Sayer 2009). In other words, using the language of justice, contributive justice imposes the obligation not to reserve a disproportional amount of meaningful work for oneself, as well as the duty to do one's share of tedious work. Doing meaningful work or tasks is valued instrumentally, as a means for self-realization and because it allows people to be less replaceable or dispensable for certain tasks, leading to wider social recognition (Timmermann and Félix 2015). Yet it is also valued intrinsically, as the mastery of complex skills is often perceived as fulfilling in itself (Sennett 2008). Here special care needs to be taken not to take current background conditions as a basis for justifying unequal divisions of labour. While meaningful tasks—such as marker-assisted plant-breeding—require a particular skillset that few people possess, and thus cannot be distributed arbitrarily, contributive justice still requires us to be critical about arrangements that end up disproportionally allocating meaningful tasks to people of certain races, gender, religious groups or social classes. Contributive justice requires us to evaluate why certain groups of people are under-represented, and to establish the appropriate educational, material and legal conditions to eliminate non-work-related hurdles and discrimination (Sayer 2009). Here we can again observe the problem of adaptive preferences, as people from historically underprivileged positions might tend to subjectively perceive as fair arrangements that would objectively be considered as unfair (Gomberg 2007; Dieterlen 2003). Instead of exploiting such "preferences", people in positions of power should educate themselves about the unjust structures that have led to this outcome and work towards redressing it.

In a strict sense, this duty also requires us to structure work environments and labour conditions so that they do not foreseeably leave specific groups of people doing less interesting tasks. For example, universities and funding agencies that depend for much of their research output on doctoral and post-doctoral researchers who are paid through scholarships or on a freelance basis without adequate labour protection rights, will systematically disadvantage mothers and those who want to fulfil their moral obligation to do their fair share of the work involved in parenting, particularly during the highly-demanding years of early childhood. The failure to account for this time spent outside academia is difficult to compensate by individual effort, and is likely to result in obtaining a position with less benefits, if any at all. As a society we pay a high price, through the loss of epistemic diversity, for establishing a research environment where committing to care work—or any form of extramural activity—incurs competitive disadvantage. Due to such pressures, the research population is unable to reflect the most basic parameters of the diversity found in the general population.

Financial rewards also affect work experience. Receiving less money than people doing the same job may erode work quality. While established customs may make such arrangements seem acceptable, transparency and comparability of tasks and remunerations make it difficult to perceive high income inequalities as fair, for example between senior and junior faculty who perform similarly in terms of teaching load and research output (Sayer 2008). Poor remunerations at an early career stage also lead to substantial class advantage, in many countries even making it practically impossible for people who are not from a rich household to professionally engage in research.

With regard to agricultural innovation, it is important to be explicit about the fact that volunteer work also falls under the scope of contributive justice. A failure to allocate the necessary resources to improve and diffuse certain types of agriculture—in particular tropical agriculture, urban agriculture and agroecology—has made such efforts largely dependent on the efforts of volunteers. While volunteer work in general may be intrinsically rewarding, this does not justify overburdening certain groups of people with tedious tasks (Sanghera 2018). Society needs to be careful not to exploit those who are more sensitive towards care needs with such social arrangements. Moreover, while the concept of volunteering implies that it is done on the basis of free will, volunteering is not a pure leisure activity and often also requires commitments. The fact that volunteer work is unpaid, or insufficiently paid, is a questionable social arrangement in itself when we as a society rely on such work for securing basic needs (Graeber 2018). The choice to offer one's agricultural expertise for free in order to solve hunger problems may be based more on knowledge of the urgency of the problem of hunger and malnutrition, than on boredom or a need to socialize. Work is often offered on a volunteer basis because it needs to be done and nobody is willing to pay for it. In other instances, volunteer work is a prerequisite or at least a major asset for advancing in certain professions or attending elite universities. Such factors need to be weighted when establishing the conditions for volunteer work, so as not to overburden people with overly demanding commitment clauses or harsh penalties, or exploit people by expecting unconditional continuity, and shaming them for non-compliance.

International scientific cooperation is often characterized by a highly unfair division of tedious and meaningful tasks. Researchers should be worried by the conscious and unconscious biases they may have, and the effect of such biases on the quality of scientific research, as well as the harm discrimination causes to fellow researchers, which may create a huge disincentive to participate in common scientific projects. Discriminating against researchers and research sources may also create epistemic poverty, as it narrows the range of resources that are subject to examination, and ignores or takes less seriously certain types of criticism and perspectives. It may also lead to the pointless duplication of research efforts. Scientists from the Global South should not be treated as second-class scholars who are only useful as assistants who perform repetitive and unchallenging tasks.

At an international level, to allow for a fairer distribution of tasks, massive efforts are needed to develop capacity building and improve local research capacities. Major differences in skills and productivity may continuously lead to the monopolization of meaningful tasks. This may come as a result of stronger research partners seizing such work for themselves, or due to the belief, which may also be supported by weaker research partners, that such a division of work is the best way to rapidly achieve a higher social goal, such as improving food security or making agriculture more ecologically sustainable. Another factor that needs to be taken into account is that stronger partners should not end up imposing conditions that are not strictly necessary for the success of a project. Sometimes standardized policies can be detrimental to the needs of collaboration partners and need to be changed to better suit local circumstances (Sangwand 2018). Stronger partners need to be aware that weaker partners may too easily accept conditions, and may feel uncomfortable demanding additions and changes. These barriers need to be broken by encouraging an open dialogue, and by recognizing each other as peers in both research and management. Even when research expenses are covered by the stronger parties, a sensible dialogue is still necessary to avoid unwanted policies being accepted by the weaker parties without any discussion as to whether those policies are actually needed (ibid.).

Here work is also needed to resist an unfair division of tasks between scientists and farmers in collaborations. An absolute emphasis on top-down knowledge transfer is not only disrespectful, but may also waste crucial reciprocal knowledge transfer opportunities between users and producers of technologies (Freire 1973). A horizontal knowledge transfer and exchange improves the quality of work, and enlarges the overall pool of available knowledge through increased cognitive diversity. It also improves technology adoption, which has been shown to be very difficult in agriculture extension (Röling 2009).

6.3.7 Recognizing People's Contributions

Recognition plays different roles. It may reveal a sense of gratitude, in the sense of acknowledging the importance of someone's contribution. An invention can be seen by others as causally contributing to a desirable outcome or improvement in overall welfare. More simply, recognition acknowledges someone's existence, independently of any awareness of their achievements or role. A common social practice is to greet people we meet on a regular basis, even if we do not know or even care exactly what they do. Another form of recognition is to formally acknowledge someone's contribution, through acknowledgments in texts or talks, citations, reviews, or even prizes.

While recognition is not needed for every valuable act people do, there can be negative effects on people's willingness and motivation to contribute to society if it is withheld from specific people or types of work. This can act as a disincentive or erode the quality of work, making it an issue of contributive justice. When a certain type of work is not recognized by society, it is less likely that people will voluntarily do it. In cases where such work is necessary, or needed for a wider project, such lack of recognition deteriorates work experience. Therefore contributive justice demands that we do not systematically fail to recognize types of work upon which we strongly rely as individuals or society, when they require comparable levels of effort or dedication. This form of justice also demands that we do not discriminate between people who perform similar tasks under similar difficulties when offering recognition. Special circumstances may also make certain people's efforts more laudable, e.g. being very young or old in age, or some forms of disability. Historical wrongs often make the accomplishments of historically disadvantaged groups particularly laudable. Here justice requires that we recognize such hurdles and absences of privilege, and acknowledge the additional difficulties that were faced.

6.4 Objections to the Idea of Contributive Justice in the Agricultural Sector

A different perspective suggests that we should forget about making agriculture more inclusive. Indeed, considerable effort within agricultural innovation is invested in making agriculture less labour-intensive. The idea behind this is that people who are freed from food production can dedicate their labour to other socially beneficial endeavours, or simply enjoy more spare time. The extent to which this direction of innovation is desirable depends on two major issues. First, does increased automation affect people's food entitlement (Mazoyer and Roudart 2006)? Second, is farming a special human interest that should be protected and allowed to be practiced in a meaningful way? Let us consider these two questions.

First, if we think that innovation should contribute to social welfare, then research should not be directed towards making people redundant, while failing to offer any concrete realizable policy proposals for amending the negative effects of such an outcome. If, by automating food production, people end up losing their food entitlement and access to the means to grow their own food, innovation will have catastrophic consequences for human welfare. There are policy proposals underway—including a reconceptualization of the nature of work—to mitigate such effects (Bottazzi 2019). One option, that is still just an idea in the political field, is to tax machinery that does human work, in order to redistribute income among those people who became unemployed due to technological advancement (Hagelücken 2017). While we could ponder many alternative proposals, the need to solve this problem is urgent (cf. Danaher 2019), as the land that is used for highly automated agriculture continuously increases.

Second, there is the question of the social value that being a farmer should have in a world where resources for producing food (particularly land and water) are of limited supply. As this is a special interest, in the sense that it is only needed for the realization of some lifestyles, we only need to make the essential goods to farm accessible. If this interest deserves social protection, as many food sovereignty advocates claim, then this would mean that society needs to conserve an innovation stream that continuously improves forms of farming that involve manual work. Ultimately, we will have to address the question of the future of farming as a democratic society, through a deliberative process involving all major stakeholders.

A very different kind of worry is the issue of contributive justice and responsibility. By opting for an inclusive research environment, and departing from a hierarchical decision-making structure, we may encounter the so-called "problem of many hands" for ethical responsibility, which may result in nobody assuming enough responsibility for the dangers of an invention or for widely diffusing the resulting innovations (Doorn 2010). These risks could be substantial if, for example, the emerging "do-it-yourself" biology movements mature (Keulartz and van den Belt 2016) and people start developing new biotechnological inventions in unregulated environments. Contributive justice demands that tedious and meaningful tasks be shared, which provides some direction when jobs that have to be done vary in their intrinsic value. Yet this duty is a poor guide when dividing tasks that are similarly interesting—such as some types of research and extension service—but which appeal to different characters differently. While this concept requires people to promote social well-being, it offers poor guidance for solving classical problems of collective action and individual responsibility.

Another issue that has been raised is that by making research projects more inclusive, and giving people with insufficient scientific training a stronger voice, we may pay a high price in terms of lower epistemic standards (Ghinea 2019). This is not a minor risk, taking into account how much time often passes until scientific mistakes are uncovered, and also that false results sometimes lead to human suffering and lost opportunities. To address this problem, we need a clear division of responsibilities within larger scientific projects that involve people with no or lower scientific qualifications.

6.5 Concluding Remarks: Facilitating Participation Through Diversification

The profit-maximizing use of intellectual property rights negatively affects people's ability to participate in agricultural innovation and its governance. This makes the maintenance of parallel innovation systems crucial for satisfying the demands of contributive justice. Parallel innovation systems can help to overcome both physical and epistemic hurdles. The diversification of innovation systems increases the chances that people will have access to food production sites, and be able to develop new spaces for producing food where previously this was not possible. For instance, rapid urbanization and the concentration of life in urban areas demands that we improve our capacity to produce food in cities. Since people generally have additional jobs, and many careers require travel, food production can only be inclusive if systems have a high degree of automation, so as to allow people to engage in parallel work (cf. Bernstein 2014). An additional option is to dedicate certain areas in cities to agroecology, and thereby build green refuge areas that provide food and ecosystem services, for which urban farmers should ideally receive some type of compensation. In rural areas, farmers can explore different forms of agriculture, depending on access to markets, environmental circumstances, regulations, sustainability commitments and directives, and cultural values and worldviews. Users of these different systems can continuously contribute towards their optimization, by sharing insights and problems via online forums and community centres.

The epistemic hurdles that smallholders face in standardized science systems make traditional ecological knowledge systems particularly attractive, as they make use of situated knowledge and are open, at the same time, to both face-to-face and online collaborations (Reyes-García et al. 2018; Walajahi 2019). Smallholders can also further improve their food systems by collaborating and exchanging with neighbours, and by making the most of locally available resources (Šūmane et al. 2018). Perhaps most importantly, ecological intensification is indispensable for allowing farmers to contribute to human welfare, as we are reaching and have surpassed dangerous thresholds in relation to our environmental footprint.

In addition to maintaining parallel innovation systems, efforts need to be made to fight discrimination and to narrow knowledge gaps, so as to make sure that people have a fair chance to participate, regardless of the social group they belong to (or are assumed to belong to). In particular, members of indigenous communities, people with disabilities and women still face major struggles in contributing to innovative projects. Much more progress needs to be made to make innovation more inclusive.

References

Altieri, Miguel A., and Parviz Koohafkan. 2008. *Enduring farms: Climate change, smallholders and traditional farming communities*. Penang: Third World Network.

Álvarez, L., and B. Coolsaet. 2020. Decolonizing environmental justice studies: A latin american perspective. *Capitalism Nature Socialism* 31 (2): 50–69. https://doi.org/10.1080/10455752.2018.1558272.

Anderson, E. 2015. Feminist epistemology and philosophy of science. In *The stanford encyclopedia of philosophy*, ed. Edward N. Zalta. Standford: Standford University.

Atenas, Javiera, and Leo Havemann. 2013. Quality assurance in the open: An evaluation of OER repositories. *INNOQUAL-International Journal for Innovation and Quality in Learning* 1 (2): 22–34.

Atenas, Javiera, Leo Havemann, and Cristian Timmermann. Forthcoming. Critical literacies for a datafied society: Academic development and curriculum design in Higher Education.

Baquedano Jer, Sandra, and Sara Larraín. 2015. The seeds of liberation in Latin America. In *Seed sovereignty, food security*, ed. Vandana Shiva, 333–344. New Delhi: Women Unlimited.

Bayruns Garcia, Eric. 2019. Expression-style exclusion. *Social Epistemology* 33 (3): 245–261.

Bernstein, Henry. 2014. Food sovereignty via the 'peasant way': A sceptical view. *Journal of Peasant Studies* 41 (6): 1031–1063. https://doi.org/10.1080/03066150.2013.852082.

Besson, Samantha. 2015. Science without borders and the boundaries of human rights. *European Journal of Human Rights* 4: 462–485.

Bornemann, Basil, and Sabine Weiland. 2019. Empowering people—democratising the food system? Exploring the democratic potential of food-related empowerment forms. *Politics and Governance* 7 (4): 105–118.

Bottazzi, Patrick. 2019. Work and social-ecological transitions: A critical review of five contrasting approaches. *Sustainability* 11 (14): 3852.

Bovenkerk, Bernice. 2012. *The biotechnology debate: Democracy in the face of intractable disagreement*. Dordrecht: Springer.

Britz, Johannes J. 2004. To know or not to know: A moral reflection on information poverty. *Journal of Information Science* 30 (3): 192–204.

Britz, Johannes J. 2008. Making the global information society good: A social justice perspective on the ethical dimensions of the global information society. *Journal of the American Society for Information Science and Technology* 59 (7): 1171–1183.

Britz, Johannes J., and Tomas A. Lipinski. 2001. Indigenous knowledge: A moral reflection on current legal concepts of intellectual property. *Libri* 51 (4): 234–246. https://doi.org/10.1515/libr.2001.234.

Britz, Johannes J., and Peter Lor. 2010. The right to be information literate: The core foundation of the knowledge society. *Innovation: Journal of Appropriate Librarianship and Information Work in Southern Africa* (11): 8–21.

Caparrós, Martín. 2014. *El hambre*. Bogotá: Planeta.

Chapman, Audrey, and Jessica Wyndham. 2013. A Human Right to Science. *Science* 340 (6138): 1291–1291. https://doi.org/10.1126/science.1233319.

COMEST. 2015. *Ethical perspective on science, technology and society: A contribution to the post-2015 agenda*. Paris: UNESCO & COMEST.

Coolsaet, Brendan. 2016. Towards an agroecology of knowledges: Recognition, cognitive justice and farmers' autonomy in France. *Journal of Rural Studies* 47: 165–171.

Danaher, John. 2019. *Automation and utopia: Human flourishing in a world without work*. Cambridge, MA: Harvard University Press.

de-Shalit, Avner. 2004. Political philosophy and empowering citizens. *Political Studies* 52 (4): 802–818.

Dieterlen, Paulette. 2003. *La pobreza: Un estudio filosófico*. México, DF: Fondo de Cultura Económica.

Donders, Yvonne. 2011. The right to enjoy the benefits of scientific progress: In search of the state obligations in relation to health. *Medicine, Health Care and Philosophy* 14: 371–381.

Doorn, Neelke. 2010. A procedural approach to distributing responsibilities in R&D networks. *Poiesis & Praxis* 7 (3): 169–188.

Editorial. 2018. Science benefits from diversity. *Nature* 558: 5–6.

Eitzel, Melissa, Jessica Cappadonna, Chris Santos-Lang, Ruth Duerr, Sarah West, Arika Virapongse, Christopher Kyba, Anne Bowser, Caren Cooper, Andrea Sforzi, Anya Metcalfe, Edward Harris, Martin Thiel, Mordechai Haklay, Lesandro Ponciano, Joseph Roche, Luigi Ceccaroni, Fraser Shilling, Daniel Dörler, Florian Heigl, Tim Kiessling, Brittany Davis, and Qijun Jiang. 2017. Citizen science terminology matters: Exploring key terms. *Citizen Science: Theory and Practice* 2 (1): 1–20. https://doi.org/10.5334/cstp.96.

Escobar, Arturo. 1998. Whose knowledge, whose nature? Biodiversity, conservation, and the political ecology of social movements. *Journal of Political Ecology* 5 (1): 53–82.

Fiske, Amelia, Lorenzo Del Savio, Barbara Prainsack, and Alena Buyx. 2019. Conceptual and Ethical Considerations for Citizen Science in Biomedicine. In *Personal Health Science*, ed. Nils B. Heyen, Sascha Dickel, and Anne Brüninghaus, 195–217. Wiesbaden: Springer.

Fraser, Nancy. 1998. Social justice in the age of identity politics: Redistribution, recognition, and participation. In *The Tanner Lectures of Human Values*, ed. Grethe B. Peterson, 1–67. Salt Lake City: The University of Utah Press.

Freire, Paulo. 1973. *¿Extensión o comunicación? La concientización en el medio rural*. México, DF: Siglo XXI.

Fricker, Miranda. 2007. *Epistemic injustice: Power and the ethics of knowing*. Oxford & New York: Oxford University Press.

Ghinea, Narcyz. 2019. Citizen science and the politicization of epistemology. *The American Journal of Bioethics* 19 (8): 58–60.

Gliessman, Stephen R. 2007. *Agroecology: The ecology of sustainable food systems*. Boca Raton: CRC Press.

Gomberg, Paul. 2007. *How to make opportunity equal*. New York: Wiley Blackwell.

Gomberg, Paul. 2016. Why distributive justice is impossible but contributive justice would work. *Science & Society* 80 (1): 31–55. https://doi.org/10.1521/siso.2016.80.1.31.

Graeber, David. 2018. *Bullshit jobs: A theory*. New York: Simon & Schuster.

Grasswick, Heidi E. (ed.). 2011. *Feminist epistemology and philosophy of science: Power in knowledge*. Dordrecht: Springer.

Gupta, Anil K. 2006. From sink to source: The Honey Bee Network documents indigenous knowledge and innovations in India. *Innovations* (summer): 49–66.

Gupta, Anil K. 2010. Grassroots green innovations for inclusive, sustainable development. In *The innovation for development report 2009–2010, strengthening innovation for the prosperity of the nations*, ed. Augusto Lopez-Claros, 137–146. Houndmills & New York: Palgrave Macmillan.

Gupta, Anil K., Anamika R Dey, Chintan Shinde, Hiranmay Mahanta, Chetan Patel, Ramesh Patel, Nirmal Sahay, Balram Sahu, P Vivekanandan, and Sundaram Verma. 2016. Theory of open inclusive innovation for reciprocal, responsive and respectful outcomes: Coping creatively with climatic and institutional risks. *Journal of Open Innovation: Technology, Market, and Complexity* 2 (1): 16.

Hagelücken, Alexander. 2017. Bill Gates fordert Robotersteuer. Süddeutsche Zeitung, February 21, 2017. https://www.sueddeutsche.de/wirtschaft/2.220/digitalisierung-bill-gates-fordert-robotersteuer-1.3386861.

Holt-Giménez, Eric, and Miguel A. Altieri. 2013. Agroecology, food sovereignty, and the new green revolution. *Agroecology and Sustainable Food Systems* 37 (1): 90–102.

Holt-Giménez, Eric, Raj Patel, and Annie Shattuck. 2010. *Rebeliones alimentarias: Crisis y hambre de justicia*. Barcelona: El Viejo Topo.

ICSU Study Group on Science and Traditional Knowledge. 2002. Science and Traditional Knowledge: Report from the ICSU Study Group on Science and Traditional Knowledge.

Jefferson, Richard A. 2006. Science as social enterprise: The CAMBIA BiOS initiative. *Innovations: Technology, Governance, Globalization* 1 (4): 13–44.

Johnson, Jeffrey Alan. 2014. From open data to information justice. *Ethics and Information Technology* 16 (4): 263–274.

Kelbessa, Workineh. 2013. Indigenous knowledge and its contribution to biodiversity conservation. *International Social Science Journal* 64 (211–212): 143–152.

Keulartz, Jozef, and Henk van den Belt. 2016. DIY-Bio–economic, epistemological and ethical implications and ambivalences. *Life Sciences, Society and Policy* 12 (1): 7.

Kitcher, Philip. 1990. The division of cognitive labor. *The Journal of Philosophy* 87 (1): 5–22.

Kloppenburg, Jack. 2005. *First the seed: The political economy of plant biotechnology*, 2nd ed. Madison: University of Wisconsin Press.

Lappé, Frances Moore, Jennifer Clapp, Molly Anderson, Robin Broad, Ellen Messer, Thomas Pogge, and Timothy Wise. 2013. How we count hunger matters. *Ethics & International Affairs* 27 (03): 251–259.

Lor, Peter Johan, and Johannes J. Britz. 2005. Knowledge production from an African perspective: International information flows and intellectual property⋆. *The International Information & Library Review* 37 (2): 61–76.

Lotter, Konrad. 2007. Die Konkordatslehrstühle an den bayerischen Universitäten. *Widerspruch* 45: 53–66.

Mazoyer, Marcel, and Laurence Roudart. 2006. *A history of world agriculture: From the neolithic age to the current crisis*. New York: Monthy Review Press.

Merton, Robert K. 1942. Science and technology in a democratic order. *Journal of Legal and Political Sociology* 1 (1–2): 115–126.

Mormina, Maru. 2019. Science, technology and innovation as social goods for development: Rethinking research capacity building from Sen's capabilities approach. *Science and Engineering Ethics* 25 (3): 671–692. https://doi.org/10.1007/s11948-018-0037-1.

Moss, Jeremy. 2002. Power and the digital divide. *Ethics and Information Technology* 4 (2): 159–165.

Ngumbi, Esther. 2019. Banish hunger on university campuses. *Nature* 574: 151.

Nyéléni Forum for Food Sovereignty. 2007. *Declaration of Nyéléni*. Sélingue: Nyéléni Forum for Food Sovereignty.s
O'Neill, John. 1993. Future generations: Present harms. *Philosophy* 68 (263): 35–51.
Olivé, León. 2013. Participación ciudadana, gestión y evaluación tecnocientífica. In *Aproximaciones a la filosofía política de la ciencia*, ed. Carlos López-Beltrán and Ambrosio Velasco Gómez, 475–488. Mexico, DF: Universidad Nacional Autonoma de Mexico.
Parr, Tom. 2015. From philanthropy to philanthropists. In *New Philanthropy and social justice: Debating the conceptual and policy discourse*, ed. Behrooz Morvaridi, 64–77. Bristol: Policy Press.
Plomer, Aurora. 2012. The right to access the benefits of science and intellectual property rights. In *Biotech innovations and fundamental rights*, ed. Roberto Bin, Sara Lorenzon, and Nicola Lucchi, 45–68. Milan: Springer-Verlag Italia.
Rawls, John. 1999. *A theory of justice*, Rev. ed. Cambridge, Mass.: Belknap Press of Harvard University Press.
Reyes-García, Victoria, Petra Benyei, and Laura Calvet-Mir. 2018. Traditional agricultural knowledge as a commons. In *Routledge handbook of food as a commons*, ed. José Luis Vivero-Pol, Tomasso Ferrando, Olivier De Schutter and Ugo Mattei, 173–184. Oxon & New York: Routledge.ss
Ribot, Jesse C., and Nancy Lee Peluso. 2003. A theory of access. *Rural Sociology* 68 (2): 153–181.
Robeyns, Ingrid. 2017. Having too much. In *Wealth: NOMOS LVI*, ed. Jack Knight, and Melissa Schwartzberg, 1–44. New York: New York University Press.
Rodney, Walter. 1972. *How Europe underdeveloped Africa*. London: Bogle-L'Ouverture Publications.
Röling, Niels. 2009. Pathways for impact: Scientists' different perspectives on agricultural innovation. *International Journal of Agricultural Sustainability* 7 (2): 83–94.
Röling, Niels G., Joseph Ascroft, and Fred Wa Chege. 1976. The diffusion of innovations and the issue of equity in rural development. *Communication Research* 3 (2): 155–170.
Rosset, Peter. 2009. Fixing our global food system: Food sovereignty and redistributive land reform. *Monthly Review* 61 (3): 114–128.
Rosset, Peter Michael, Braulio Machin Sosa, Adilén María Roque. Jaime, and Dana Rocío Ávila Lozano. 2011. The Campesino-to-Campesino agroecology movement of ANAP in Cuba: Social process methodology in the construction of sustainable peasant agriculture and food sovereignty. *The Journal of Peasant Studies* 38 (1): 161–191.
Sanghera, Balihar. 2018. Contributive injustice and unequal division of labour in the voluntary sector. *Sociological Research Online* 23 (2): 308–327.
Sangwand, T-Kay. 2018. Preservation is political: Enacting contributive justice and decolonizing transnational archival collaborations. *KULA: Knowledge Creation, Dissemination, and Preservation Studies* 2 (1).
Sayer, Andrew. 2008. Moral economic regulation in organizations: A university example. *Organization* 15 (2): 147–164.
Sayer, Andrew. 2009. Contributive justice and meaningful work. *Res Publica* 15: 1–16. https://doi.org/10.1007/s11158-008-9077-8.
Sayer, Andrew. 2012. Capabilities, contributive injustice and unequal divisions of labour. *Journal of Human Development and Capabilities* 13 (4): 580–596. https://doi.org/10.1080/19452829.2012.693069.
Schoonmaker, Sara. 2007. Globalization from below: Free software and alternatives to neoliberalism. *Development and Change* 38 (6): 999–1020.
Sen, Amartya. 1981. *Poverty and famines. An essay on entitlement and deprivation*. Oxford & New York: Oxford University Press.
Sennett, Richard. 2008. *The craftsman*. New Haven & London: Yale University Press.
Shaver, Lea. 2010. The right to science and culture. *Wisconsin Law Review* 121: 121–184.
Shiva, Vandana. 2009. Women and the gendered politics of food. *Philosophical Topics* 37 (2): 17–32.

Shue, Henry. 1996. *Basic rights: Subsistance, affluence, and U.S. foreign policy*, 2nd ed. Princeton, N.J.: Princeton University Press.

Singer, Peter. 2009. *The life you can save: How to do your part to end world poverty*. New York: Random House.

Sparrow, Robert, and Robert E. Goodin. 2001. The competition of ideas: Market or garden? *Critical Review of International Social and Political Philosophy* 4 (2): 45–58. https://doi.org/10.1080/13698230108403349.

Steinvorth, Ulrich. 2009. The right to work and the right to develop one's capabilities. *Analyse & Kritik* 1: 101–113.

Šūmane, Sandra, Ilona Kunda, Karlheinz Knickel, Agnes Strauss, Talis Tisenkopfs, Ignacio des Ios Rios, Maria Rivera, Tzruya Chebach, and Amit Ashkenazy. 2018. Local and farmers' knowledge matters! How integrating informal and formal knowledge enhances sustainable and resilient agriculture. *Journal of Rural Studies* 59: 232–241.

Taylor, Robert. 2004. Self-realization and the priority of fair equality of opportunity. *Journal of Moral Philosophy* 1 (3): 333–347. https://doi.org/10.1177/1740468104001000307.

Thompson, Paul B. 2017. *The spirit of the soil: Agriculture and environmental ethics*. New York: Routledge.

Timmermann, Cristian. 2014. Sharing in or benefiting from scientific advancement? *Science and Engineering Ethics* 20 (1): 111–133.

Timmermann, Cristian. 2018. Contributive justice: An exploration of a wider provision of meaningful work. *Social Justice Research* 31 (1): 85–111. https://doi.org/10.1007/s11211-017-0293-2.

Timmermann, Cristian. 2019. Citizen science for biomedical research and contributive justice. *The American Journal of Bioethics* 19 (8): 60–62.

Timmermann, Cristian, and Georges F. Félix. 2015. Agroecology as a vehicle for contributive justice. *Agriculture and Human Values* 32 (3): 523–538.

Timmermann, Cristian, Georges F. Félix, and Pablo Tittonell. 2018. Food sovereignty and consumer sovereignty: Two antagonistic goals? *Agroecology and Sustainable Food Systems* 42 (3): 274–298. https://doi.org/10.1080/21683565.2017.1359807.

Tittonell, Pablo. 2014. Ecological intensification of agriculture—sustainable by nature. *Current Opinion in Environmental Sustainability* 8: 53–61.

Tittonell, Pablo. 2019. Las transiciones agroecológicas: Múltiples escalas, niveles y desafíos. *Revista de la Facultad de Ciencias Agrarias* 51 (1): 231–246.

Tittonell, Pablo, Laurens Klerkx, Frederic Baudron, Georges F Félix, Andrea Ruggia, Dirk van Apeldoorn, Santiago Dogliotti, Paul Mapfumo, and Walter AH Rossing. 2016. Ecological intensification: local innovation to address global challenges. *Sustainable Agriculture Reviews* 19: 1–34

Torrance, Andrew W., and Eric von Hippel. 2015. The right to innovate. *Michigan State Law Review* 2015 (2): 793–829.

Vayena, Effy, and John Tasioulas. 2015. "We the Scientists": A human right to citizen science. *Philosophy & Technology* 28 (3): 479–485.

Via Campesina. 1996. *The right to produce and access land*. Rome: Via Campesina.

Walajahi, Hina. 2019. Engaging the "Citizen" in citizen science: Who's actually included? *The American Journal of Bioethics* 19 (8): 31–33.

Wiggins, Andrea, and John Wilbanks. 2019. The rise of citizen science in health and biomedical research. *The American Journal of Bioethics* 19 (8): 3–14.

Young, Iris Marion. 2004. Responsibility and global labor justice. *Journal of Political Philosophy* 12 (4): 365–388. https://doi.org/10.1111/j.1467-9760.2004.00205.x.

Ziegler, Jean. 2011a. *Destruction massive: Géopolitique de la faim*. Paris: Seuil.

Ziegler, Rafael. 2011b. Fair competition of ideas. In Responsible Innovation Conference, The Hague, April 18–19, 2011.

Chapter 7
Intergenerational Justice and Innovation for Long-Term Agricultural Sustainability

Abstract Innovation in agriculture brings about a number of positive and negative externalities. In this chapter I will focus on one particular externality, which is how innovation affects the consumption of non-renewable or slowly renewable resources that are essential for securing the human right to food in the future. The reproduction and massive use of some of these innovations require the right to destroy the effectiveness of resources that were not created by the inventor nor those buying the inventions. The use of pesticides leads to a loss of their effectiveness due to biological resistance, leading to the destruction of the resource (i.e. of its effectiveness) and genetic pollution (i.e. an increase in resistant biological organisms). Similarly, the use of high-yield crop varieties leads to the loss of soil fertility. While the destruction of these resources is inevitable when using these inventions, there are a number of measures that can be taken to prolong the active life of these resources. Users of these resources can comply with strict usage regulations and exclusive rights holders can assert substantial pressure to make sure users generally comply with such policies.

Keywords Future generations · Conservation farming · Responsible resource use · Soil erosion · Pesticides · Reciprocity · Common heritage

7.1 Introduction: The Long-Term Effects of Agricultural Innovation

Innovation may have long-term effects on people and nature. Depending on the targets it sets and the innovation paths it follows, these effects may be positive, neutral or detrimental for future people and the environment. A primary aim of agriculture is to produce food in large enough quantities to feed the population. Throughout history, agricultural innovation has concentrated on increasing yields and on reducing fluctuations between different years. The latter was of vital importance in the past, due to limited ability to build up stocks, protect them from predators and spoilage, and prevent pillage. While capacity to stockpile food has increased significantly, particularly among richer countries, stable yields are still crucial for regions that need or want to rely on their own food production. Short-term fluctuations in yields

C. Timmermann, *Social Justice and Agricultural Innovation*, The International Library of Environmental, Agricultural and Food Ethics 31,
https://doi.org/10.1007/978-3-030-56193-2_7

and rapid impoverishment of soils have often been paid for with hunger, particularly among poorer countries and regions who cannot buy imported food in global markets. This problem, together with increased vulnerabilities to climate change, calls for improving not only yields but also the resilience of food systems (Funes-Monzote 2008).

Historically, the negative consequences of poor soil and pest management were seen at different rates, sometimes within a few years, and at other times in a matter of decades (Mazoyer and Roudart 2006). As a people's attachment to a specific geographic region grew, be it for political, cultural or religious reasons, the need to be able to maintain and increase harvest yields increased. Sustainability became a necessity, in order to stay in the same place and not be forced to move to new lands. As populations grew and land became increasingly scarce, sustainable agricultural practices became not only a means to stay in a specific place, but increasingly an issue of justice. In a world of scarce resources, specifically land and water, the rapid exhaustion of resources would irremediably lead to fights over resources, often escalating into wars (McMichael 2017).

In many places, nature itself was able to restore the resources needed for food production over time through longer periods of fallowing. In other cases, soils could be recovered through techniques that accelerate natural processes, for example by attracting termites which play a key role in regenerating soil by untightening its structure (Félix et al. 2018). In other places, massive deforestation led to irreversible soil erosion and changes of climate, or even to desertification. For example, massive deforestation during the Roman times in the Eastern Mediterranean changed the landscape so drastically that the loss of valuable agricultural land has still not been recovered (Mazoyer and Roudart 2006).

Dependence on slowly renewable or non-renewable agricultural resources is relatively old, particularly for fertilizers. Early modern Europe was dependent on the clearing of enormous areas of forests to produce potash. Later, in the industrial age, guano sediments and saltpetre were imported in large quantities from the Peruvian coast and the Chilean north, to treat European eroded soils (Travis 2015). With the development of synthetic fertilizers, agriculture became dependant on cheap fossil fuels to maintain soil fertility (Patel and Moore 2017).

Nowadays, the large amounts of agrochemicals used in agriculture have led to new types of damage that need much more time to remediate, if this can be done at all. The effects of unsustainable food practices will also be increasingly felt beyond our present generation. We will therefore continue our analysis by moving on to another category of social justice: intergenerational justice.

In this chapter, we will discuss the problems raised by innovation when it triggers or accelerates the use of exhaustible resources, such as pesticide effectiveness (Timmermann 2015) or phosphorus (de Ponti et al. 2012), or when it leads to the loss of slowly renewable resources, such as soil fertility (El Mujtar et al. 2019, Puig de la Bellacasa 2015) and agrobiodiversity (Shiva and Pandey 2006), in relation to the rights and interests of future generations. When research and development lead to marketable goods, it may happen that these innovations draw from an exhaustible pool of resources: something that has been amply studied in the case of antibiotic resistance (Viens and Littmann 2015; Outterson 2005, 2014). In relation to agriculture, the use of inventions that exhaust non-renewable or slowly renewable resources may make future food production much more difficult, or even jeopardize the ability of upcoming generations to feed themselves altogether. At the same time, innovation can play a key role in restoring degraded and polluted resources, such as land and water, and may lead to more sustainable alternatives for producing food, through ecological intensification or digital farming. Justice demands that we also acknowledge both the positive and negative effects on future generations' well-being that innovation may have, and examine if the balance of costs and benefits is at least hypothetically acceptable.

The strong need for these types of innovation makes their enclosure through intellectual property rights and their insufficient diffusion particularly worrying. By working with theories of intergenerational justice, we can construct an argument for constraining the irrational use of these resources and redirecting research towards the aims of sustainability.

7.2 On the Use and Destruction of Resources

Innovations in agriculture bring about a number of positive and negative externalities. While, for example, improving methods of capturing carbon in agricultural land have substantial benefits for food security and countering climate change (Altieri et al. 2015), the use of a number of agrochemicals had to be prohibited after harmful effects on both human and animal life were proved (Kroma and Flora 2003). It is hard to assess the long-term impact that the massive use of agricultural innovation will have on people and the environment, as well as the consequences of failing to develop or make use of innovations (McIntyre et al. 2009). Here I focus on two kinds of externalities: first, on how innovation affects the consumption of non-renewable or slowly renewable resources that are essential for securing the human right to food in the future and, second, on how innovation directly and indirectly affects future opportunities to live in a harmonious relationship with nature, which is a fundamental interest and a requirement for many forms of cultural self-realization and worldviews (Kelbessa 2015; Rozzi 2013).

It is inevitable that humanity, especially considering our numbers, ends up using resources that are of limited supply. The key question, however, is not whether we should use these resources at all. To leave resources unused or underused is also a form of wastage. As John Locke already recognized, we inherited plenty so that we do not have to suffer from scarcity (Locke 1689). What is at stake is ensuring that available resources are consumed responsibly, and that we leave "enough and as good" for others—and here "others" includes, as I will argue here, also future generations (Wolf 1995). This calls both for efficient use and for providing adequate substitution and compensation for the depletion of resources. To support this, we need to argue for the responsible use of resources as an issue of justice. The new realities brought about by climate change, biodiversity loss and the massive erosion of our soils, reveal that due to our current population size and its overall impact on the environment, we cannot rely on traditional senses of responsibility grounded in close interpersonal relationships between generations, or on identifying with the interests of one's descendants, to guarantee a responsible use of resources and the establishment of conservation measures that accord with the foreseeable interests of future generations. While, in a normative dimension, there is much hesitance to argue that future generations have rights in any strict sense, it is clear that they have a strong claim on the natural resources needed to grow food to secure their future basic needs, and also a strong interest in these goods that should not be ignored in favour of present banal cravings (Brock 1998).

Generally speaking, theories of intergenerational justice do not demand that current generations make major sacrifices so that future generations will have minor improvements in their lives (Meyer 2016). These theories, however, usually embrace principles of fairness and long-term sustainability. For example, principles of intergenerational reciprocity demand that we do not make life for future generations more difficult than what we ourselves had to endure. Similarly, concerning the ability to enjoy nature, we should not destroy natural environments unless we have compelling reasons to do so, as people from future generations will have a strong interest in enjoying these landscapes. Therefore, we proceed to analyse both incentives to use resources efficiently, and thereby reduce the environmental impact of our food production systems, and incentives to adequately compensate future people for the resources that we have used (Beekman 2004).

The availability of resources in the future depends not only on what we leave for future generations, but also on the demand for resources in the future. The technologies we develop now will be crucial for ensuring that future generations can live comfortably, while at the same time using resources more efficiently. Yet another major factor that will influence demand is population size. Our failure to halt sudden sharp population increases in the present may have enormous consequences on the welfare of future generations (Meijers 2017). To avoid any rational Malthusian fears, humanity needs to develop its technological capacity and knowledge at a rate that is appropriate for solving the social and environmental challenges it will face. Society needs to develop such solutions and build infrastructure at a rate that is proportional to population growth, so as to avoid overburdening future generations. However, even then, we may stipulate that surpassing a certain population size on the planet will make it impossible to live in harmony and in connection with nature. Moreover, from a moral perspective, a world with a larger human population is not necessarily a better world. Unfortunately, to avoid a major deviation from our central topic, I must refrain from discussing the issue of population in more detail.

When looking at how intellectual property regimes incentivize efficient use, we find several problems when market incentives are not properly aligned to efficiency criteria and optimal diffusion. In the case of non-proprietary science and communal farming systems, we may find similar inefficiencies or wasteful behaviour when social condemnation fails to discourage the wasteful use of resources and people stubbornly avoid switching to more environmentally sound methods, such as using mulching instead of slash-and-burn techniques to regenerate soils (Tomich et al. 2011; Félix 2019). We will proceed by analysing these incentives and social norms.

7.3 Intellectual Property, Market Incentives and Efficient Use

As discussed in Chap. 2, nowadays a substantial amount of research in the life sciences is done for profit, and is incentivized by offering intellectual property rights for inventions that meet legally predefined requirements. Due to the rights traditionally associated with property, if we want to argue for efficient use, we need to start by asking ourselves to what extent people *own*, as innovators and consumers, the invention they have developed or acquired.

We may recall, that traditionally—that is, according to Lockean property theories—ownership of the fruits of one's labour is justified, as researchers mix their work with something that is essentially unowned and available in abundance. In theory, recognizing property rights allows people to benefit from the fruits of their labour and encourages the improvement of property, something that is ultimately to the benefit of all (Strauss 1952). This reasoning has been translated, in the realm of intellectual property, into a *right to exclude* others from using the invention in question. Here we need to keep in mind that, as with property rights for tangible objects (Honoré 1961), intellectual property rights consist in a so-called "bundle of rights", which may or may not incorporate a series of sub-rights (Faraci and Jaworski 2014). Besides having a right to exclude others, it is of special relevance to find out whether owners of innovations also have the right to use the invention, the right to destroy the invention, and the right to transfer these rights (that is, to alienate the invention).

The first problem we encounter is that intellectual property rights do not unconditionally deliver a right to use the invention. An exclusive right is, first of all, a right to exclude others from using the invention. An innovator may have the right to exclude others while not being entitled to use the resource herself (Liivak and Peñalver 2013). Governments retain the power to forbid use. As mentioned before, public interests and biosafety concerns may prohibit innovators from using certain inventions even after being awarded a patent. Generally, unless it is subject to specific regulations, innovators can assume that they can use their invention provided that such usage does not cause harm or annoyance.

Second, even when a right of use has been conceded (or when there is no opposition to the exercise of this right), we still have to distinguish between being entitled to use a good and having the right to destroy a good (Strahilevitz 2005). The latter, in the case of intangibles, can have substantial consequences. If we interpret such a right as the right to destroy a single copy of an invention, usually not much harm follows. However, if we interpret such a right as the right to destroy a type of invention along with the tangible element which it uses, we may encounter far greater issues for justice. For example, if I own a landrace that includes unique crop varieties with important identified traits, I would be destroying something of important social value if I were allowed to exercise full property rights and thus decided to burn fields without making sure that those varieties are properly conserved. Public interests may trump the exercise of some property rights, for example by depriving owners of the right to destroy irreplaceable assets (Goodin 1983).

The reproduction and massive use of some of these innovations requires the right to destroy the effectiveness of resources that were not created by the inventor nor by those buying the inventions. The use of pesticides leads to a loss of their effectiveness due to biological resistance, leading to the destruction of the resource (i.e. its effectiveness) and genetic pollution (i.e. an increase in resistant biological organisms) (Timmermann 2015). For example, we can observe a major conflict of interests between biotechnological seed corporations and organic farmers in relation to maintaining the long-term pesticidal efficacy of *bacillus thuringiensis* (Bt). Currently *bacillus thuringiensis* is an open access resource that can be freely exploited. Therefore the biotechnology industry can patent genetically modified varieties with Bt genes inserted, accelerating the resistance of pathogens to the natural pesticide and ultimately leaving organic farmers without any effective pesticide (van den Belt 2015). The lack of common agreement on how this resource should be governed is leading to a rapid loss in its efficacy (Frisvold and Reeves 2010).

Similarly, intensive use of monocrops, repeated application of agrochemicals and soil tillage reduce the amount of soil bacteria, and extract minerals at a rate that makes full soil restoration impossible (Tittonell 2016; Mazoyer and Roudart 2006). In extreme cases this leads to the loss of soil fertility: something that may take years or even decades to restore (El Mujtar et al. 2019). Market pressures may encourage the use of improved seed varieties in ecologically irresponsible ways, as illustrated by the extremely high proportion of monocrops (over 90%) in Latin American soy fields (Leguizamón 2014).

Users of inventions cannot simply assume that they have the right to destroy goods that they have not brought into existence while exploiting their developed or acquired invention. While the destruction of such resources is inevitable when using these inventions at a large scale, there are a number of measures that can be taken to prolong the active life of these resources. Users of these resources can comply with strict usage regulations, and exclusive rights' holders can assert substantial pressure to make sure users generally comply with such policies through contractual agreements (Robaey 2016b). There have been instances where commercial seeds producers have used this power to oblige seed users to establish refugee zones, in order to slow down resistance or maintain a certain ratio of crop diversity, so as to increase the resilience of fields and reduce the impact of erosion (Noonan 2003). Even if society does not oppose the gradual destruction of such resources, governments and seed companies have to work hand in hand to establish adequate incentives to facilitate the sustainable use of such goods (Frisvold and Reeves 2010), in accordance with the potential interests of future people.

As for creating an incentive for the sustainable stewardship of pesticide effectiveness, a market-friendly solution would be to extend the patent protection period.

While they have an exclusive right over the invention, patent holders have an incentive to see that their invention continues to be marketable, which first of all requires that it does not excessively decline in efficacy during this period. Translating this to pesticides, patent holders will have a commercial interest in customers' establishing refuge areas that slow down the prevalence of organisms that are resistant to the pesticide by offering a habitat for competing organisms and natural predators. This type of solution has been criticized for enclosing something that should be part of the public domain and thus undermining the very justification of patents, which is to secure a contribution to publicly available knowledge in return for temporary exclusivity (Timmermann 2015; Outterson 2005). Such a method of incentivizing the maintenance of pesticide effectiveness excludes poorer potential consumers who cannot pay the higher fees that usually follow from monopolies.

Third, we are left with the question of whether innovators are allowed to transfer any of these rights. This question is crucial not only in relation to the privileges that intellectual property grants, but most importantly for society, in relation to identifying who assumes the responsibilities that arise concerning innovation (Robaey 2016a). Consecutive transfers from one owner to another should not have the outcome that issues of responsibility end up unaddressed. It is crucial to have clarity about who assumes responsibility, as this will help us to identify who is liable to offer compensations.

In order to encourage efficient use, society needs to regain awareness of which property rights are indeed transferred to innovators and which rights are retained. As this brief overview has shown, innovators often assume that they can use whatever other goods come their way when grasping the benefits of the fruits of their labour. The granting of rights to the use of resources that innovators have not brought into existence has, however, to be negotiated with society, taking into account the interests of both present and future people.

Furthermore, it must be pointed out that intellectual property rights create a strong incentive to develop inventions that have a limited life-time. If an invention becomes obsolete, then people are obliged to continuously purchase protected inventions, as the inventions that fall in the public domain will have little utility. This business strategy has been called "planned obsolescence", or "accelerated obsolescence", and is having disastrous effects on our natural environment (Guiltinan 2009). Strictly speaking, planned obsolescence is not necessarily linked to innovation, and can of course exist independently, as sometimes it merely involves manufacture. Yet in more extreme cases, innovation can be tightly linked to planned obsolescence by assisting in the development of materials that will break shortly after legally mandated warranties expire, or when products are made to last only as long as similar products developed by business competitors. Here we need to add the risk of illegal collusions in maintaining such practices, which are particularly difficult to control when the sector is dominated by a few big players, as is the case with the seed and agrochemical industry. Another strategy is to create an artificial need for vulnerable add-ons for important tools that may make the whole tool useless once it breaks down. For example, farm machinery may have sophisticated additional functions that stall the whole machine if it becomes broken or blocked. Such practices have disastrous consequences for sustainability, as they involve massive wastage in the manufacture, transportation and disposal of tangible innovations. They also raise issues of global justice, since inaccessible rural areas in the Global South have major difficulties in accessing replacement parts and technical service providers (Ziegler 2011).

Lastly, intellectual property facilitates industry expansion, and thereby encourages the development of innovations that can be sold at a large scale or at high prices. Innovations can affect sustainability in ways that are not straightforward. Nowadays, it can be seen that farms of over 1,000 ha are those with the largest post-harvest losses (Ricciardi et al. 2018). Since large tractors that give their owners major advantages can be only acquired by richer large-scale land owners (Mazoyer and Roudart 2006), innovation in agricultural machinery thereby gives users who produce massive food waste a significant advantage. This effect ultimately has an enormous impact on environmental sustainability, in terms of water use and contamination, greenhouse gas contributions, and soil erosion.

7.4 On the Efficient Use of Non-proprietary Innovation

In relation to efficient use, the choice of leaving a resource outside the proprietary system may lead to diverse outcomes. First of all, the outcome will depend on whether the resource is seen as belonging to nobody (*res nullius*) or to a community (*res communis*). If we adopt Hardin's (1968) interpretation of the commons as an area that is not governed by either formal or informal laws or rules, we are likely to observe a well-known consequence that is depicted metaphorically as a pasture that becomes overgrazed as a result of farmers adding too many cattle. While individuals gain by increasing their consumption of resources from the commons, the community as a whole benefits from constraint, that is, by accepting usage restrictions (Aguilera Klink 1991; Ostrom 1990). Conversely, if people perceive the resource as belonging to humanity as a whole, and to both current and future generations, in the sense of the human community, then at least in theory we should expect a different pattern of behaviour that favours efficient use and the provision of compensatory public goods. Yet, such behaviour mostly depends on people's goodwill, and is difficult to enforce unless non-compliance is sanctioned with social stigma. Ideally, the erosion of such common goods should be seen as an aggregate harm that users should take collective action to prevent, for example, by establishing and adhering to regulations (cf. Kahn 2014). While recognizing a resource as a common heritage of humankind may provide a weak incentive for conservation, the provision of compensatory public goods involves massive costs which people will usually avoid paying unless they are coerced, or have committed to doing so by a common agreement.

In the case of agrobiodiversity, we need to remind ourselves that conservation is not promoted by restraint, but rather by the actual use of a diversified pool of crop resources (Halewood 2013; Batur and Dedeurwaerdere 2014). This requires farmers to incorporate a diverse pool for conservation, even when this does not meet present economic interests. In this case, conservation is achieved by diversifying usage and maintaining certain thresholds, instead of abstinence.

The way in which a resource is exploited will most likely not vary much if it the invention was accessible after the expiration of intellectual property rights, or comes from incremental innovation or open science. As long as exploitation is allowed without restriction, the way an innovation was brought into existence will have little effect on promoting sustainability. With the important exception of local heritage crop varieties, users rarely have a clear notion of the effort involved in developing resources and their uniqueness. Therefore, it would be insufficient to assign the responsibility for conserving these resources to their users alone, and this calls for states to assume responsibility for establishing and enforcing regulation that would secure their future availability.

7.5 Justifying Responsible Use and Compensation: Innovating for Long-Term Sustainability

In the political sphere, the view that finds the greatest consensus is the principle of sustainable development. Here the question to be asked is whether an innovation system, in relation to its impact on the environment and the use of exhaustible resources, meets the interests of future generations. The best-known definition of this principle is attributed to the Brundtland report (1987): "sustainable development is development that meets the needs of the present without compromising the ability of future generations to meet their own needs." Traditionally, this definition has been interpreted as not jeopardizing the ability of future generations to meet such basic needs as food and healthcare, and often also the crucial resources needed to sustain democratic institutions and shelter, such as energy, communication infrastructure and construction materials. Nowadays increased scientific knowledge and environmental changes have made it clear that future generations may not only have an *interest* in conserving natural habitats, but actually have a strong *need* for the conservation of nature, due to the ecological services it delivers and on which human life depends (IPBES 2019). By framing nature as *the* means to cover fundamental needs, such as food and health, future generations assert a strong claim on the conservation of nature. This does not suggest that we should set aside any claims that future generations have that are interest-based, i.e. claims that are grounded in values or mere aesthetic appreciation (Brock 1998). Such claims exert a strong demand to restructure and redesign our production systems, especially our food systems, due to their enormous impact on nature. Let us review the interest-based and need-based arguments that demand a transition towards sustainability.

The moment we, as a global society, want to abide by the principle of sustainable development, it becomes necessary that we adapt our current policies and practices to live by this principle. In the case of conventional agriculture, we can observe a number of practices and methods that lead to massive environmental deterioration. Here we need to remind ourselves that industrial agriculture is highly dependent on fossil fuels for the manufacture of fertilizers, and exhausts phosphorus reserves and erodes soils (Tittonell 2013). In order to align our behaviour to the principle of sustainable development, we need to change how we produce food and make major adjustments to our diets, particularly by reducing the amount of animal products we consume (Singer 2004; Korthals 2004). Here we need to cut our consumption of products with a high environmental footprint and reduce waste in all stages of food production, delivery and consumption (Gjerris and Gaiani 2013). As urban populations grow, we must pay special attention to developing incentives to reintegrate unused organic materials back into the soils, as they are an important input for soil fertility (Timmermann et al. 2018). In addition, we need to replace production systems and diets with more sustainable alternatives. This will require the development of alternative farming systems, such as ecological intensification, and of arguments to encourage people to adopt diets and food customs with a lower environmental impact.

Increasingly, there is a strong consensus that sustainability has not only an ecological but also social and economic dimensions (Werkheiser and Piso 2015). Agricultural innovation that develops systems that increasingly rely on external inputs places farmers in a position of strong dependency. Although dependency is not wrong in itself, it makes farmers vulnerable, especially when interactions occur between highly unequal parties or it is not bidirectional (Ervin et al. 2010). Farmers are vulnerable to currency devaluation, which make it harder to acquire inputs produced abroad. They are also vulnerable to changes in the sales strategies of input producers: risking losses in self-determination and being exposed to exploitative price increases. Farmers are also vulnerable to geopolitical changes that may impede access to imported goods, such as embargos, blockades or war (Altieri et al. 1999). For many areas, natural disasters may also impede the timely delivery of needed agricultural inputs. Moreover, since fertilizers are dependent on cheap oil, rises in oil prices lead to increases in fertilizer prices (Patel and Moore 2017). This dependence on externally produced inputs reduces the resilience of agricultural systems (Funes-Monzote 2008). Here we must recall that food is not like any other commodity, but is rather a basic good. Innovation that leads to uncontrolled dependence is socially unsustainable, particularly when methods to produce their own inputs are thereby lost. It also limits the options for future generations when they come to decide whether they really want to continue towards further industrialization or instead switch to ecological practices and principles of de-growth.

The choice to live by the principle of sustainable development can be legitimized by a democratic process. Here, for example, people could decide to establish the public good "sustainable development" by acknowledging its tangible and intangible benefits for present and future generations. Among the benefits of establishing such a public good, we can list the moral advantages of living in a world that respects the foreseeable interests of future generations: something that is even more pressing now, when some of the costs of our food production system will have to be paid by people who are already born. In addition, we improve the chances that future generations will be able to continue to guard and enjoy our material and immaterial heritage, as well as further develop some of today's projects. It also improves stability and respect between generations, particularly that had by the younger for the older: something that we are in increased need of, as shown by the school protests from all over the world that we are witnessing at the time of this writing. Due to issues of costs, and our inability to shield our borders from ecological interactions, sustainable development requires global cooperation (Kelbessa 2014). In other words, sustainable development can only be sustained in the long-term if it is recognized and established as a global public good. Nonetheless, principles of sustainable development need to be adhered to at all levels, from individual to community to regional networks. The establishment of this good also requires the implementation of various policies and institutions, as well as the development of new technologies and social innovations. The costs of such a transition and its maintenance would need to be fairly distributed, and the social recognition of sustainable development as a public good would allow us to condemn non-contributing society members as free-riders. Yet here it is important to recognize that the unwillingness to cooperate of a few should not legitimize others in not doing their fair share. It has even been argued that when one is able to contribute more than one's fair share towards alleviating the problems of the needy, and others are not doing their part, one is obliged to do more than the agreed minimum. The injustice of having to do more than one's fair share (while still living in plenitude) is no moral excuse for avoiding one's obligation to help others in need (Stemplowska 2016). If not enough is done for the sustainable development of agriculture, then people in the future will not be able to meet their needs.

7.6 What Can Theories of Intergenerational Justice Contribute to the Discussion?

What happens if people, especially those outside one's community, do not abide by such a principle? To start, it is important to highlight that the current rate of environmental destruction is largely avoidable. We already possess a wide range of technologies and methods for considerably reducing our carbon footprint. We know, for example, how to compost food waste, thereby capturing methane that can be used as fuel instead of contributing to climate change, and use the residues to assist soil fertilization, instead of using exclusively synthetic fertilizers that require large amounts of fossil fuels for their manufacture (Thi et al. 2015; Tittonell 2013). These innovations are largely underused. We also have research networks and methods for developing new innovations: an infrastructure that can be widely and quickly expanded when needed or desired, as history has shown with the unfortunate example of scientific research in war efforts. This experience explains why massive proposals that are nowadays under discussion to address the current environmental crisis, such as the Green New Deal, have wide support and are perceived as achievable (Klein 2019). Furthermore, we not only have the scientific capabilities to develop technological innovations, but also have the scientific capacity to develop a wide range of social innovations to stimulate the use of green and greener technologies, and improve people's cooperation in reducing our environmental footprint, e.g. by increasing the attractiveness of composting and transforming human habitats into shared refuge areas for nature. Considering both the current and potential room for action, environmental deterioration—at least as we experience it today—is largely an avoidable harm.

As a second issue, there are the costs involved in avoiding harming people in the future. With the exception of reducing air pollution from cars in large urban agglomerations, in the mainstream media we hear little about the present-day benefits of a sustainability transition. While some areas of the world are undertaking quite expensive infrastructure changes to reduce their impact on nature, other areas have hardly adopted any green technologies at all. Especially in relation to agriculture, some countries have improved productivity at the cost of an enormous impact on the natural environment, while others have hardly adopted any innovations, thus producing insufficient food and relying on imports (Oteros-Rozas et al. 2019). Ecological intensification could help to improve yields in those areas of the world where they are lowest, thus contributing to food security and sustainability by improving soils and increasing biomass (Tittonell et al. 2016): something that will also mitigate climate change by capturing carbon. The biodiversity on which agroecological farms are dependent also leads to diversified diets, which have immediate health benefits (Altieri 2003). Our environmental impact can be greatly reduced without making any major sacrifice in terms of current well-being. Quite the contrary, pollution is already a major public health threat. Moreover, access to green areas has mental health benefits (Bratman et al. 2019). The immediate benefits delivered by a transition to sustainable development should be fully acknowledged.

To continue, as a third issue, we must assess the strength of the claim on conserving natural environments. As mentioned before, future generations have a strong need for the conservation of nature due to the ecological services it delivers, which are essential for securing their basic needs. Yet, as we know from everyday experience, this does not mean that we need all of nature to be conserved, but "merely" much of it. The limits of what we can safely destroy or alter in nature are highly debatable—we may have already reached a tipping point with our carbon emissions and their impact on anthropogenic climate change (McMichael 2017). We can only estimate what can be fixed through technology. Besides the mere human survival argument, future generations are very likely to have a strong interest in nature conservation for aesthetic and animal rights reasons. From a wider social rights perspective, we must keep in mind that living in mutualistic relationships with nature is important for many indigenous communities and their cultural understanding, as well as for many worldviews.

There are a number of different lines of reasoning used to justify the demands of intergenerational justice (Gosseries and Meyer 2009; Tremmel 2006). Let us analyse five lines of argument used to justify these rights, so as to gain some insight into the diversity of arguments for defending the interests of future generations: reciprocity-based arguments, the idea of a common heritage of humankind, generational fair shares, generational sovereignty, and rights-based claims.

Reciprocity. Perhaps the most straightforward and intuitive line of argument for defending the interests of future generations involves a call for reciprocity. According to reciprocity principles, we should offer future generations the same opportunities to live a flourishing life that we ourselves have received from previous generations (Gosseries 2009). Reciprocity in this sense can mean that we were awarded a number of services (e.g. care) by the preceding generation that we cannot reciprocate in the same way to them, but we can reciprocate these services by assisting the next generation (Weston 2012). This would be in the interest of the previous generation as well, since fulfilling long-term plans such as having a heritage and achieving the continuity of cultural, environmental and scientific projects, requires successors (O'Neill 1993).

This does not usually mean that future generations should have the exact same opportunities that we have, but that they at least ought to have a comparable set of opportunities and burdens. Since previous generations provided us with a set of opportunities, we should reciprocate this effort and constraint by also giving equally valuable opportunities "back" to the next generation (Dennis 2015). If we consume exhaustible resources, justice demands that we offer some type of good in compensation, ideally in the form of a public good, so as to make sure that the compensation is not illicitly appropriated by more powerful groups in the future. For example, providing environmentally useful knowledge as compensation for the consumption of fossil fuels could at least initially count as a compensatory public good. Moreover, the good offered to future generations should also foreseeably be perceived by future people as adequate compensation. This provides a strong argument for offering innovations to save energy and improve ecologically sustainable production as compensation. It also sets a limit to how far we can offer banal compensatory goods for the loss of the last remaining exemplars of a species, since the extinction of a species can rarely be adequately compensated, as there is hardly any human-made good that might count as an adequate substitute. Here we could use a hypothetical "envy test", and ask ourselves if future generations may envy our present set of opportunities in comparison to what they are likely to have (cf. Weston 2012). Reciprocity is not only concerned with close future generations; current generations should also not engage in practices that will have strongly detrimental consequences in the far future.

Common heritage. Another possibility is that people reach a formal agreement indicating that a resource should be the common property of present and future generations. A prominent example of this type of agreement, in favour of conserving resources for future generations, is the principle of the common heritage of humankind (Chemillier-Gendreau 2002; Shackelford 2008). According to this principle, there are a number of goods that are passed between generations on the condition that the recipients steward them and continue to pass them on. These goods are meant to enrich our lives, yet this common heritage comes also with obligations (Wolfrum 1983). First, the principle demands common management of these goods. Second, the benefits that any group or individual extract from such goods, including scientific knowledge, should be shared with the rest of humanity. And third, this principle rejects any kind of harmful use of such goods, specifically condemning military uses. Nowadays the concept is widely applied to specified regions, such as international waters, the seabed, certain exceptional natural habitats (while retaining national sovereignty rights) and parts of the Earth's orbit, and even to outer space. The concept has also been adopted for humanmade artefacts, such as key historical objects, works of art and written heritage.

While the principle of common heritage of humankind is well-known in international organizations, in particular UNESCO (Turtinen 2000), there is seldom any mention of how each generation acquires its obligations in relation to this inherited heritage. It seems that each generation receives the goods that make up part of our common heritage as a closed package, and by accepting this complete package each generation not only commits to conserving those goods that are appreciated at that moment, but also those goods which immediately display little utility or value. Heritage comes with burdens and benefits, and to be true to its nature, it should not consist of an opportunistic selection of inherited goods based on immediate needs. As with history, a selective remembrance, or in our case, a selection of heritage items for conservation, would deform our historical identity, as it only reveals one side of the past. We would distort our history by omitting parts of it. This is particularly worrying due to the strong ties that many indigenous people have to native species as tokens of their heritage and elements of their cultural practices (Rozzi 2013).

This framework is particularly useful for justifying the conservation of traditional knowledge of agricultural practices, especially those that have been time-tested and have a strong sustainability potential. It also encourages the protection of heritage varieties. By identifying genetic resources and traditional knowledge as heritage, we accept an obligation to conserve not only what is momentarily useful for us in the present, but also what could be valued for the sake of cultural diversity and for potential future use. As heritage, these resources are valued beyond their instrumental value. Once an agreement is reached to formally acknowledge such resources as common heritage of humankind, policies can be developed and enforced to significantly extend the availability and conservation of these resources. The use (or non-use) of a good that is considered common heritage is then regulated by specific sets of rules that are decided on a case-by-case basis (Timmermann and Robaey 2016). The explicit reference to (hu)mankind as the heir of this heritage underlines the importance of this good beyond borders and generations (Taylor 2013).

The massive commercialization of seeds has, however, pushed this approach away from political discussions, and the early Undertaking on Plant Genetic Resources has largely been discarded in favour of access and benefit sharing approaches (Louwaars 2012; de Goede 2014; Mgbeoji 2003).

Fair shares. Another argument for limiting the consumption of exhaustible or slowly renewable resources is to say that each generation is entitled to something like a generational fair share. Yet, when used to guide policy-making, the notion of a generational fair share stumbles against major limitations. To start, if we are going to divide existing exhaustible resources into fair shares, we can either base our calculations on previous consumption (something that has been shown to be unsustainable), or opt for a most likely arbitrary end point, and divide the remaining exhaustible resources into a series of shares which each generation may dispose of. This requires us to be able to quantify the amount of resources left on the planet with sufficient precision for intergenerational distributive purposes. In the latter case, it would be particularly problematic to socially accept the premise that there is a distant end point—however far away this end may be—in relation to which we are dividing resources: something that is hardly feasible as a policy guideline, since we largely live our lives and understand our generation on the assumption of there being a future humanity. Furthermore, implementing generational shares would ideally demand a complete social consensus on stabilizing population numbers: something that we are still far from reaching. Any increase in population, especially if it is sudden, would force future generations to live with much less per capita: something that may not always be compensated by efficiency improvements achieved through technological innovation. Lastly, if the resources we consume are to be understood as distributive goods, then we should also have some idea of what they are worth now and in the future, which would require some idea of their potential use. Yet, in relation to genetic resources and traditional knowledge, we are losing both at an alarming rate, before we have managed to get even a vague idea of their potential use and social value (Robinson 2008).

Generational sovereignty. Ideally each generation should live with a degree of independence from decisions made by previous generations. This would allow something that could be described as "generational sovereignty" (Gosseries 2016). In contrast, if we spend more resources than we are entitled to, we will constrain too extensively the options available in the future, limiting future generations' autonomy. To redress past injustices in terms of resource depletion, some generation might end up being obliged to renounce significant liberties, in order to restore the functioning of crucial ecosystems and thereby permit the existence of future generations. We can already witness some of these major undertakings: perhaps the massive reforestation efforts in various parts of the world are the most publicized examples. A broader understanding of generational sovereignty will also condemn certain technological path dependencies and policies that imperil food sovereignty. If some generation loses the means for its members to grow their own food as independent smallholders, then it loses a fundamental skill needed to maintain the independence of future generations.

Rights-based claims. Finally, in international law, the most widely accepted arguments are rights-based, particularly those based on human rights. As long as there is a common wish to procreate and continue human civilization, we are obliged by socially agreed human rights commitments not to obstruct future generations' abilities to secure their basic needs (Kumar 2009), especially their urgent and continuous need for food. Even if some people do not wish to procreate, they still need to take into consideration that they are relying on other people who continue to have children, so that someone will be there to take care of them when they are in need: in old age or after any serious accident (Olsaretti 2013). Moreover, we also count on the existence of a future generation for fulfilling present-day interests, as many projects require that someone will carry them out and maintain them in the future, as is the case with scientific advancement (Kumar 2018). It would be deluded to pretend that the current generation is completely indifferent to what future generations will do with what they will inherit (O'Neill 1993).

Generally speaking, in order for there to be a future, we need to conserve enough of nature to allow the delivery of the needed ecosystem services and permit the production and delivery of food. The unfortunate rapid destruction of these resources, however, results in a substantial normative convenience. The negative consequences of the destruction of these resources will already affect the ability of people who are now children to secure the coverage of basic needs for their offspring. Since having children that can live a flourishing life is a substantial interest of people who already exist, but are legally disenfranchised from asserting political pressure, we do not face the same normative problems as we do when defending the rights of future people directly, such as the famous non-identity problem (Zakaras 2016). This allows us to formulate a stronger set of moral demands, backed by a human rights framework that will have a much wider positive reception.

Taking these considerations together, we can highlight two main factors:

(1) Having children is a widely asserted human interest.
(2) If we have children, justice demands that they have at least the same opportunities to live a flourishing life as their parents' generation.

This would provide a strong basis for agreeing on the conservation of resources, the development of more sustainable food production methods, and switching to diets with a lower environmental impact.

7.7 What Policy Recommendations Can We Draw from These Arguments?

Securing the interests of future generations calls for strong compliance in the efficient use of resources and the protection of natural habitats. At first sight, this would require avoiding wastage: something that often already benefits people in the present. Taking efficient use more seriously would, however, also require developing technologies to enable much more efficient use of available resources, as seen in other areas of innovation, such as the development of energy-saving light bulbs and information technologies. Developing such technologies and diffusing them may involve substantial costs, yet they are necessary for the sustenance of future human populations (Barry 1997).

Yet, as mentioned before, it is unavoidable that certain resources will be exhausted. In this respect, the most obvious solution for addressing the interests of future generations is to provide a compensatory good, ideally to prevent future injustices in the form of a public good, since goods of this type do not exclude people, and can simultaneously be enjoyed by those who want to make use of them. Knowledge can be a very suitable compensatory public good. For example, the loss of the effectiveness of a pesticide could be compensated by innovations to reduce the need for applying pesticides, such as integrated pest management systems. Similarly, the loss of soil

Table 7.1 Three proposals for compensating future generations

Policy proposal	Main problem	Realizability
(1) Irreplaceability	We cannot seek consent from future people	– Limits innovation – Very demanding
(2) Technological fix	We have equal claims and should offer compensation	– Innovation-friendly – Low demand
(3) Technological fix + compensatory good	We have equal claims, but we should not leave future people in a position of disadvantage with inadequate compensation	– Innovation-friendly – Demanding

quality could be compensated by knowledge of how to speed up the recovery of soils (Tittonell 2016; Félix 2019). Clearly, such bargains with future generations have limitations and cannot be indiscriminately offered to compensate for the loss of unique resources (Zwarthoed 2016). The loss of species, for example, is difficult to substitute, especially in a future world where little biological diversity is left. Here it is crucial that, at least hypothetically, future generations would agree to the terms of such intergenerational bargains. A problem here is that we can only make rough guesses about how future generations will value, intrinsically and instrumentally, a resource that we have exhausted and the public good provided as compensation (cf. Bovenkerk 2015). The extinction of species has been and continues to be a paradigmatic example.

Taking what we have discussed until now into consideration, we can think of three tentative policy proposals for alleviating the negative effects of the destruction of these essential resources (Table 7.1). The first and most demanding proposal consists of strict liability for eventual damages. The second proposal is to oblige users and producers to develop technological innovations that will be able to fix the damages caused by the use of earlier inventions. The third and most modest proposal demands the creation of public goods of sufficient magnitude to qualify as adequate compensation. Let us briefly sketch and evaluate these options from a normative perspective.

With the first option, we could question entirely our current practice in which we do not fully internalize the negative externalities of our innovation systems. According to a strict liability approach, future generations should not inherit any unwanted burdens from previous generations. We would not be free to choose possible compensations, since we are not able to seek their consent to find out if they would really agree to such substitutions. This position would be considerably restrictive and would halt much scientific advancement. Here the idea that we could compensate future generations while consciously harming them is rejected, in favour of replacing research trajectories and technology use with options that do not conflict with the major foreseeable interests of future generations.

The second option is the least burdensome for present generations, since it would only require that we offer something as compensation, in the sense of "this for that" or technological fixes for problems we have caused. This allows quite a lot of room to cheat by offering simple or merely symbolic compensation if this demand is taken lightly, but can in principle also be interpreted as imposing an obligation to offer something of similar value in return for the exhausted resource. This proposal assumes that it is morally acceptable to exhaust a resource that future generations may have an interest in, as long as one offers compensation. Technological fixes also tempt to keep fixing a problematic established system instead of deciding for more sustainable alternatives (Scott 2011).

The third option is somewhat more demanding. Since we cannot know how future generations will value the resources that we have consumed, and it is highly plausible that they will set a higher value on them than we have, we must not only offer compensation, but also a bonus to compensate any unexpected value that we fail to identify. This is important, for the obvious reason that future generations cannot demand goods from us after we have ceased to exist. Here it is recognized that both present and future generations may have an interest in a resource, and that it is acceptable to exhaust the resource while using it efficiently and offering adequate compensation.

There is a more general problem that needs to be discussed, which is that, especially in the natural world, compensations cannot substitute lost goods. Let us now turn to this problem.

7.8 On Rejecting the Substitution of Lost Nature or Seed Diversity

To some, the very idea of compensating while harming or destroying nature sounds preposterous. The cost that we have paid, in terms of human suffering and natural destruction during the last six centuries of massive expansion of capitalist production forms, is devastating (Patel and Moore 2017). Many indigenous worldviews and philosophical perspectives condemn the parasitic relationships that we as a species have developed with the natural environment, and demand the opportunity to live in alternative relations that are more mutualistic with other living beings. We find such perspectives in African ubuntu ethics (Kelbessa 2014), in Buddhism (Suh 2014), and in many Latin American native worldviews, such as honoring Pachamama and related conceptions (Gudynas 2011; Waldmueller and Rodríguez 2018), as well as in many contemporary positions in environmental ethics and political philosophy (Nussbaum 2006).

A strong rejection of continuing to destroy natural habitats would require a massive change in our food systems and an obligation to drastically reduce our environmental footprint. In practice, this means that we will have to give up the idea that we are morally authorized to consume exhaustible resources (i.e. destroy nature) as long as we offer adequate compensation. If we recognize a strong interest in, or even a need for nature in future generations, we would be harming them with our present destruction of nature. The idea that we can continue to harm while compensating is usually rejected when it comes to interpersonal relations, with the exception of cases where consent was obtained and no violation of human dignity is at stake. In some sense, offering a compensatory good for the destruction of nature is like being able to pay for its destruction: something that could even imply that one is morally legitimized to do so (cf. Gosseries 2015). Using this analogy, we could employ standard arguments to the effect that some goods (i.e. the freedom to destroy nature) should not be up for sale (Satz 2010; Sandel 2012). The liberty to destroy an irreplaceable good, to which future generations may have a legitimate claim, should not be for sale. Such an approach does not mean that we will have a standstill in innovation. Once the idea that we should work with nature and not against it becomes established, we will open up a whole different world of innovation pathways (e.g. biomimicry, see Blok and Gremmen 2016; Stojanovic 2019).

7.9 Concluding Remarks

To avoid destroying the natural environment and to secure the basic food needs of people in the future, we need policies that stimulate users of these resources to adhere to strict compliance regulations, and we need companies to use their powers to enforce these policies through responsible technology licensing. In case such policies fall short, we also need to secure a provision of public goods that will counter the negative effects of destroying these essential resources.

We saw that there is a very strong case for improving the sustainability of food systems in order to do justice to future generations. However, most views do not encourage the use of a specific type of food system. As long as adequate food is made available without jeopardizing future food production, many positions are neutral about how food is to be produced. This could be achieved through ecological intensification, or by relying much more on information technology and biotechnology (digital farming), so that we improve our abilities to grow food in cities. There are different possible future food scenarios. We could forget about farming as a special (or even protected) way of life, give agrarian land back to nature, and grow food in special hermetic towers in the city. Alternatively, we could transform cities and peri-urban areas into food producing areas, where people and semi-domesticated animals live side by side, thus incorporating much more nature into our surroundings (Altieri et al. 1999).

Additional work is needed to identify more arguments in support of a food system that allows future democratic control, as interest in having more control over food systems is on the rise: something that we can currently observe with the food sovereignty movement (Nyéléni Forum for Food Sovereignty 2007). Sustainability arguments have concentrated on environmental issues, leaving matters of social sustainability largely aside. Here it will be crucial to link intergenerational justice arguments not only with the securement of basic needs, but also with guaranteeing the human right to adequate food, which specifically allows people to object to food on non-nutritional grounds and determine their own food policies. Much as farmers' organizations are demanding nowadays in the food sovereignty movement, future generations may have an interest in maintaining democratic control over the means of food production.

References

Aguilera Klink, Federico. 1991. ¿La tragedia de la propiedad común o la tragedia de la malinterpretación en economía? *Agricultura y Sociedad* 61: 157–183.

Altieri, Miguel A. 2003. Dimensiones éticas de la crítica agroecológica a la biotecnología agrícola. *Acta Bioethica* 9 (1): 47–61.

Altieri, Miguel A., Nelso Companioni, Kristina Cañizares, Catherine Murphy, Peter Rosset, Martin Bourque, and Clara I. Nicholls. 1999. The greening of the "barrios": Urban agriculture for food security in Cuba. *Agriculture and Human Values* 16 (2): 131–140.

Altieri, Miguel A, Clara I Nicholls, Alejandro Henao, and Marcos A Lana. 2015. Agroecology and the design of climate change-resilient farming systems. *Agronomy for Sustainable Development* 1–22.

Barry, Brian. 1997. Sustainability and intergenerational justice. *Theoria* 44 (89): 43–64.

Batur, Fulya, and Tom Dedeurwaerdere. 2014. The use of agrobiodiversity for plant improvement and the intellectual property paradigm: Institutional fit and legal tools for mass selection, conventional and molecular plant breeding. *Life Sciences, Society and Policy* 10: 14.

Beekman, Volkert. 2004. Sustainable development and future generations. *Journal of Agricultural and Environmental Ethics* 17 (1): 3–22.

Blok, Vincent, and Bart Gremmen. 2016. Ecological innovation: Biomimicry as a new way of thinking and acting ecologically. *Journal of Agricultural and Environmental Ethics* 29: 203–217.

Bovenkerk, Bernice. 2015. Public deliberation and the inclusion of future generations. *Jurisprudence* 6 (3): 496–515.

Bratman, Gregory N., Christopher B. Anderson, Marc G. Berman, Bobby Cochran, Sjerp de Vries, Jon Flanders, Carl Folke, Howard Frumkin, James J. Gross, Terry Hartig, Peter H. Kahn, Ming Kuo, Joshua J. Lawler, Phillip S. Levin, Therese Lindahl, Andreas Meyer-Lindenberg, Richard Mitchell, Zhiyun Ouyang, Jenny Roe, Lynn Scarlett, Jeffrey R. Smith, Matilda van den Bosch, Benedict W. Wheeler, Mathew P. White, Hua Zheng, and Gretchen C. Daily. 2019. Nature and mental health: An ecosystem service perspective. *Science Advances* 5 (7): eaax0903. https://doi.org/10.1126/sciadv.aax0903.

Brock, Gillian. 1998. Future generations, natural resources, and property rights. *Ethics and the Environment* 119–130.

Brundtland, Gro Harlem. 1987. *Report of the world commission on environment and development: Our common future*. New York: United Nations.

Chemillier-Gendreau, Monique. 2002. The idea of the common heritage of humankind and its political uses. *Constellations* 9 (3): 375–389.

de Goede, Ludo. 2014. Global justice and the shift in property regime for plant genetic resources. *Asian Biotechnology and Development Review* 16 (1): 35–67.

de Ponti, Tomek, Bert Rijk, and Martin K. van Ittersum. 2012. The crop yield gap between organic and conventional agriculture. *Agricultural Systems* 108: 1–9.

Dennis, Dan. 2015. Property rights, future generations and the destruction and degradation of natural resources. *Moral Philosophy and Politics* 2 (1): 107–139.

El Mujtar, Verónica, Nacira Muñoz, Barbara Prack Mc Cormick, Mirjam Pulleman, and Pablo Tittonell. 2019. Role and management of soil biodiversity for food security and nutrition; where do we stand? *Global Food Security* 20: 132–144.

Ervin, David E., Leland L. Glenna, and Raymond A. Jussaume. 2010. Are biotechnology and sustainable agriculture compatible? *Renewable Agriculture and Food Systems* 25 (2): 143–157.
Faraci, David, and Peter Martin Jaworski. 2014. To inspect and make safe: On the morally responsible liability of property owners. *Ethical Theory and Moral Practice* 17 (4): 697–709.
Félix, Georges F. 2019. Slash-and-mulch: Exploring the role of shrub-based agroforestry systems for smallholder farmers in the Sahel. Ph.D., Farming systems ecology, Wageningen University.
Félix, Georges F., Ibrahima Diedhiou, Marie Le Garff, Cristian Timmermann, Cathy Clermont-Dauphin, Laurent Cournac, Jeroen CJ. Groot, and Pablo Tittonell. 2018. Use and management of biodiversity by smallholder farmers in semi-arid West Africa. *Global Food Security* 18: 76–85.
Frisvold, George B., and Jeanne M. Reeves. 2010. Resistance management and sustainable use of agricultural biotechnology. *AgBioForum* 13 (4): 343–353.
Funes-Monzote, Fernando. 2008. Farming like we're here to stay: The mixed farming alternative for Cuba. Ph.D. thesis, Wagenigen University.
Gjerris, Mickey, and Silvia Gaiani. 2013. Household food waste in Nordic countries: Estimations and ethical implications. *Etikk I Praksis-Nordic Journal of Applied Ethics* 7 (1): 6–23.
Goodin, Robert E. 1983. The ethics of destroying irreplaceable assets. *International Journal of Environmental Studies* 21: 55–66.
Gosseries, Axel. 2009. Three models of intergenerational reciprocity. In *Intergenerational Justice*, ed. Lukas H. Meyer and Axel Gosseries, 119–146. Oxford.
Gosseries, Axel. 2015. What's wrong with trading emission rights. In *Climate change and justice*, ed. Jeremy Moss, 89–106. Cambridge: Cambridge University Press.
Gosseries, Axel. 2016. Generational sovereignty. In *Institutions for future generations*, ed. Iñigo González-Ricoy and Axel Gosseries, 98–113. Oxford: Oxford University Press.
Gosseries, Axel, and Lukas H. Meyer (eds.). 2009. *Intergenerational justice*. Oxford and New York: Oxford University Press.
Gudynas, Eduardo. 2011. Buen Vivir: Today's tomorrow. *Development* 54 (4): 441–447.
Guiltinan, Joseph. 2009. Creative destruction and destructive creations: Environmental ethics and planned obsolescence. *Journal of Business Ethics* 89 (1): 19–28.
Halewood, Michael. 2013. What kind of goods are plant genetic resources for food and agriculture? Towards the identification and development of a new global commons. *International Journal of the Commons* 7 (2): 278–312.
Hardin, Garrett. 1968. The tragedy of the commons. *Science* 162 (3859): 1243–1248.
Honoré, Anthony M. 1961. Ownership. In *Oxford essays in jurisprudence*, ed. A. Guest, 107–147. London: Oxford University Press.
IPBES. 2019. *Summary for policymakers of the global assessment report on biodiversity and ecosystem services of the intergovernmental science-policy platform on biodiversity and ecosystem services*. Bonn: IPBES Secretariat.
Kahn, Elizabeth. 2014. The tragedy of the commons as an essentially aggregative harm. *Journal of Applied Philosophy* 31 (3): 223–236.
Kelbessa, Workineh. 2014. Can African environmental ethics contribute to environmental policy in Africa? *Environmental Ethics* 36: 31–61.
Kelbessa, Workineh. 2015. African environmental ethics, indigenous knowledge, and environmental challenges. *Environmental Ethics* 37 (4): 387–410.
Klein, Naomi. 2019. *On fire: The (burning) case for a green new deal*. New York: Simon & Schuster.
Korthals, Michiel. 2004. *Before dinner: Philosophy and ethics of food*. Dordrecht: Springer.
Kroma, Margaret M., and Cornelia Butler Flora. 2003. Greening pesticides: A historical analysis of the social construction of farm chemical advertisements. *Agriculture and Human Values* 20 (1): 21–35.
Kumar, Rahul. 2009. Wronging future people: A contractualist proposal. In *Intergenerational justice*, ed. Lukas H. Meyer and Axel Gosseries, 251–272. Oxford: Oxford University Press.
Kumar, Rahul. 2018. Future generations. In *Oxford Handbook of distributive justice*, ed. Serena Olsaretti, 689–710. Oxford: Oxford University Press.

Leguizamón, Amalia. 2014. Modifying Argentina: GM Soy and socio-environmental change. *Geoforum* 53: 149–160.

Liivak, Oscar, and Eduardo M. Peñalver. 2013. The right not to use in property and patent law. *Cornell Law Review* 98: 1437–1493.

Locke, John. 1689. *Two treatises of government*, ed. Peter Laslett. Cambridge: Cambridge University Press.

Louwaars, Niels. 2012. Seed science in the 21st century: Rights that scientists have to deal with. *Seed Science Research* 22 (S1): S9–S14.

Mazoyer, Marcel, and Laurence Roudart. 2006. *A history of world agriculture: From the neolithic age to the current crisis*. New York: Monthy Review Press.

McIntyre, Beverly D., Hans R. Herren, Judi Wakhungu, and Robert T. Watson. 2009. *International assessment of agricultural knowledge, science and technology for development (IAASTD): Synthesis report with executive summary: A synthesis of the global and sub-global IAASTD reports*. Washington, DC: Island Press.

McMichael, Anthony. 2017. *Climate change and the health of nations: Famines, fevers, and the fate of populations*. Oxford: Oxford University Press.

Meijers, Tim. 2017. Citizens in appropriate numbers: Evaluating five claims about justice and population size. *Canadian Journal of Philosophy* 47 (2–3): 246–268.

Meyer, Lukas H. 2016. Intergenerational justice. In *Stanford Encyclopedia of philosophy*, ed. Edward N. Zalta.

Mgbeoji, Ikechi. 2003. Beyond rhetoric: State sovereignty, common concern, and the inapplicability of the common heritage concept to plant genetic resources. *Leiden Journal of International Law* 16: 821–837.

Noonan, Douglas S. 2003. An economic model of a genetic resistance commons: Effects of market structure applied to biotechnology in agriculture. In *Battling resistance to antibiotics and pesticides: An economic approach*, ed. Ramanan Laxminarayan, 263–287. Washington, DC: Resources for the Future.

Nussbaum, Martha C. 2006. *Frontiers of justice : Disability, nationality, species membership, the Tanner lectures on human values*. Cambridge, MA: The Belknap Press, Harvard University Press.

Nyéléni Forum for Food Sovereignty. 2007. *Declaration of Nyéléni*. Sélingue: Nyéléni Forum for Food Sovereignty.

O'Neill, John. 1993. Future generations: Present harms. *Philosophy* 68 (263): 35–51.

Olsaretti, Serena. 2013. Children as public goods? *Philosophy & Public Affairs* 41 (3): 226–258.

Ostrom, Elinor. 1990. *Governing the commons: The evolution of institutions for collective action*. Cambridge: Cambridge University Press.

Oteros-Rozas, Elisa, Adriana Ruiz-Almeida, Mateo Aguado, José A. González, and Marta G. Rivera-Ferre. 2019. A social–ecological analysis of the global agrifood system. *Proceedings of the National Academy of Sciences*. https://doi.org/10.1073/pnas.1912710116.

Outterson, Kevin. 2005. The vanishing public domain: Antibiotic resistance, pharmaceutical innovation and intellectual property law. *University of Pittsburgh Law Review* 67: 67–123.

Outterson, Kevin. 2014. New business models for sustainable antibiotics. In *Centre on Global health Security Working Group Papers, Working Groups on Antimicrobial Resistance, Paper 1*. London: Chatham House (The Royal Institute of International Affairs).

Patel, Raj, and Jason W. Moore. 2017. *A history of the world in seven cheap things: A guide to capitalism, nature, and the future of the planet*. Oakland, CA: University of California Press.

Puig de la Bellacasa, Maria. 2015. Making time for soil: Technoscientific futurity and the pace of care. *Social Studies of Science* 45 (5): 691–716.

Ricciardi, Vincent, Navin Ramankutty, Zia Mehrabi, Larissa Jarvis, and Brenton Chookolingo. 2018. How much of the world's food do smallholders produce? *Global Food Security* 17: 64–72.

Robaey, Zoë. 2016a. Gone with the wind: Conceiving of moral responsibility in the case of GMO contamination. *Science and Engineering Ethics* 22 (3): 889–906.

Robaey, Zoë. 2016b. Transferring moral responsibility for technological hazards: The case of GMOs in agriculture. *Journal of Agricultural and Environmental Ethics* 29 (5): 767–786.

Robinson, Daniel. 2008. Beyond 'protection': Promoting traditional knowledge systems in Thailand. In *Patenting lives: Life patents, culture and development*, ed. Johanna Gibson, 121–138. Aldershot: Ashgate.

Rozzi, Ricardo. 2013. Biocultural ethics: From biocultural homogenization toward biocultural conservation. In *Linking ecology and ethics for a changing world: Values, philosophy, and action*, ed. Ricardo Rozzi, S.T.A. Pickett, Clare Palmer, Juan J. Armesto, and J. Baird Callicott, 9–32. Dordrecht: Springer.

Sandel, Michael J. 2012. *What money can't buy: The moral limits of markets*. New York: Farrar, Straus and Giroux.

Satz, Debra. 2010. *Why some things should not be for sale: The moral limits of markets*. Oxford: Oxford University Press.

Scott, Dane. 2011. The technological fix criticisms and the agricultural biotechnology debate. *Journal of Agricultural and Environmental Ethics* 24 (3): 207–226.

Shackelford, Scott J. 2008. The tragedy of the common heritage of mankind. *Stanford Environmental Law Journal* 27: 101–157.

Shiva, Vandana, and Poonam Pandey. 2006. *Biodiversity based organic farming: A new paradigm for food security and food safety*. New Dehli: Navdanya.

Singer, Peter. 2004. *One world: the ethics of globalization*, 2nd ed. New Haven and London: Yale University Press.

Stemplowska, Zofia. 2016. Doing more than one's fair share. *Critical Review of International Social and Political Philosophy* 19 (5): 591–608.

Stojanovic, Milutin. 2019. Biomimicry in agriculture: Is the ecological system-design model the future agricultural paradigm? *Journal of Agricultural and Environmental Ethics* 32 (5–6): 789–804.

Strahilevitz, Lior Jacob. 2005. The right to destroy. *The Yale Law Journal* 114: 781–854.

Strauss, Leo. 1952. On Locke's doctrine of natural right. *The Philosophical Review* 61 (4): 475–502.

Suh, Jungho. 2014. Towards sustainable agricultural stewardship: Evolution and future directions of the permaculture concept. *Environmental Values* 23 (1): 75–98.

Taylor, Prue. 2013. The future of the common heritage of mankind. In *Confronting ecological and economic collapse: Ecological integrity for law, policy and human rights*, ed. Laura Westra, Prue Taylor, and Agnès Michelot, 32–46. Oxon: Routledge.

Thi, Ngoc Bao, Gopalakrishnan Kumar Dung, and Chiu-Yue Lin. 2015. An overview of food waste management in developing countries: Current status and future perspective. *Journal of Environmental Management* 157: 220–229.

Timmermann, Cristian. 2015. Pesticides and the patent bargain. *Journal of Agricultural and Environmental Ethics* 28 (1): 1–19. https://doi.org/10.1007/s10806-014-9515-x.

Timmermann, Cristian, Georges F. Félix, and Pablo Tittonell. 2018. Food sovereignty and consumer sovereignty: Two antagonistic goals? *Agroecology and Sustainable Food Systems* 42 (3): 274–298. https://doi.org/10.1080/21683565.2017.1359807.

Timmermann, Cristian, and Zoë Robaey. 2016. Agrobiodiversität, das Gemeinschaftserbe-Prinzip und Marktanreize. In *Biopatente – Saatgut als Ware und als öffentliches Gut*, ed. Barbara Brandl and Stephan Schleissing, 109–132. Baden-Baden: Nomos.

Tittonell, Pablo. 2013. *Farming systems ecology: Towards ecological intensification of world agriculture*. Wageningen: Wageningen Universiteit.

Tittonell, Pablo. 2016. Feeding the world with soil science: Embracing sustainability, complexity and uncertainty. *Soil Discuss*. https://doi.org/10.5194/soil-2016-7.

Tittonell, Pablo, Laurens Klerkx, Frederic Baudron, Georges F. Félix, Andrea Ruggia, Dirk van Apeldoorn, Santiago Dogliotti, Paul Mapfumo, and Walter AH. Rossing. 2016. Ecological intensification: Local innovation to address global challenges. *Sustainable Agriculture Reviews* 19: 1–34.

Tomich, Thomas P., Sonja Brodt, Howard Ferris, Ryan Galt, William R. Horwath, Ermias Kebreab, Johan HJ. Leveau, Daniel Liptzin, Mark Lubell, and Pierre Merel. 2011. Agroecology: A review from a global-change perspective. *Annual Review of Environment and Resources* 36: 193–222.

Travis, Anthony S. 2015. *The synthetic nitrogen industry in World War I: Its emergence and expansion*. Cham: Springer.

Tremmel, Joerg Chet (ed.). 2006. *Handbook of intergenerational justice*. Cheltenham: Edward Elgar.

Turtinen, Jan. 2000. *Globalising heritage: On UNESCO and the transnational construction of a world heritage*. Stockholm: Stockholm Center for Organizational Research.

van den Belt, Henk. 2015. Design for values in agricultural biotechnology. In *Handbook of ethics, values, and technological design: sources, theory, values and application domains*, ed. Jeroen van den Hoven, Pieter E. Vermaas, and Ibo van de Poel, 571–588. Dordrecht: Springer.

Viens, A.M., and Jasper Littmann. 2015. Is antimicrobial resistance a slowly emerging disaster? *Public Health Ethics* 8 (3): 255–265.

Waldmueller, Johannes, and Laura Rodríguez. 2018. Buen Vivir and the Rights of Nature. In *Routledge Handbook of development ethics*, ed. Jay Drydyk and Lori Keleher, 234–247. London and New York: Routledge.

Werkheiser, Ian, and Zachary Piso. 2015. People work to sustain systems: A framework for understanding sustainability. *Journal of Water Resources Planning and Management* 141 (12): A4015002.

Weston, Burns H. 2012. The theoretical foundations of intergenerational ecological justice: An overview. *Human Rights Quarterly* 34 (1): 251–266.

Wolf, Clark. 1995. Contemporary property rights, Lockean provisos, and the interests of future generations. *Ethics* 105 (4): 791–818.

Wolfrum, R.üdiger. 1983. The principle of the common heritage of mankind. *Heidelberg Journal of International Law* 43: 312–337.

Zakaras, Alex. 2016. Democracy, children, and the environment: A case for commons trusts. *Critical Review of International Social and Political Philosophy* 19 (2): 141–162.

Ziegler, Jean. 2011. *Destruction massive: Géopolitique de la faim*. Paris: Seuil.

Zwarthoed, Danielle. 2016. Should future generations be content with plastic trees and singing electronic birds? *Journal of Agricultural and Environmental Ethics* 29 (2): 219–236.

Chapter 8
Procedural Justice and the Realization of Just Outcomes in Agricultural Innovation

Abstract Conceptions of procedural justice have great potential to help improve the design of processes and guidelines meant to improve social and environmental outcomes in agricultural innovation. In this chapter I introduce the general principles of procedural justice applied to agricultural innovation and discuss some of the specific procedures concerning good scientific practice and the incorporation of genetic resources and traditional knowledge. Informed consent and inclusive decision-making procedures are defended as the central demands of procedural justice.

Keywords Fair process · Corruption · Democratic decision-making · Power imbalances · Realization of social justice

8.1 From the Identification of Social Justice Issues to the Realization of Justice

As we saw, agricultural innovation has profound effects on both society and the environment, affecting multiple forms of social justice. To address the demands of social justice it becomes imperative to have fair processes to stimulate positive innovation outcomes, and to reduce to a minimum any negative effects of innovation, for people now and in the future, at home and abroad. This requires us to identify the basic principles that allow people to engage in fair processes to prevent injustices from coming about, and to restore relations after any unforeseen or remaining injustices or wrongs that could not be avoided. The former is the goal of this chapter and the latter the aim of the next chapter.

C. Timmermann, *Social Justice and Agricultural Innovation*, The International Library of Environmental, Agricultural and Food Ethics 31,
https://doi.org/10.1007/978-3-030-56193-2_8

8.2 Applying Principles Procedural Justice to Agricultural Innovation

In this chapter we will explore the potential of procedural justice to reach outcomes that are good for the environment and for present and future people. In general terms, procedural justice refers to how procedures should be structured so that they are likely to lead to fair outcomes (Doorn 2010). By establishing a set of principles and guidelines, procedural justice seeks to achieve balance between the aim of building a system that leads to accurate results, and not overburdening people with a too costly system consisting in an unworkably complex apparatus of rules and disproportionately expensive truth-seeking mechanisms (Solum 2004). The major contribution of procedural justice to agricultural innovation is that its principles improve the overall fairness of innovation systems by insisting on the dignified, respectful and polite treatment of people, thus allowing people to feel like valuable members of the community by giving them a voice and a chance to change their circumstances (cf. Vermunt and Steensma 2016). Thus procedural justice also helps to develop guidelines to facilitate a respectful debate in which different people can express themselves and where diverse arguments can be assessed, and thereby secure their right to self-determination and comply with their duty to obtain prior, free and informed consent where applicable.

Procedural justice imposes a series of demands on agricultural innovation, some of which are also justified using the other forms of justice discussed in the preceding chapters. These are:

- To create guidelines and discussion forums to ensure that research agendas plan already in the innovation proposal and design phases for wider accessibility of innovation efforts, so that the public can benefit from scientific advancement. This demand reduces the problems that will have to be faced by working with principles of justice in exchange (Chap. 4).
- To institute decision-making mechanisms that will allow research to be directed to respond to social and environmental needs according to their urgency, and divest from research that has detrimental effects on humans and nature. This demand is also partly backed by applying the concept of distributive justice defended in Chap. 5 to direct research agendas toward social needs and, when issues of sustainability are at stake, also by notions of intergenerational justice (Chap. 7).
- To increase the participation of stakeholders in technology assessment and governance, as well as the drafting of governmental science agendas and incentives (Doorn 2010). People who participate in such discussions should be treated fairly and encouraged to share their observations and opinions. This demand is also supported by contributive justice (Chap. 6), although this latter concept goes much further, as it demands that people have the opportunity and a certain obligation to develop capacities to meaningfully participate in such discourses.

- To allow production systems that are in line with principles of intergenerational justice (Chap. 7), such as ecological intensification, we need to ensure that innovation can be adequately embedded in an ecosystem and community, so as to facilitate mutually supportive relations (Gliessman 2007).
- To establish systems to prevent scientific misconduct, discrimination and harassment. Restorative measures and their justifications to repair damaged relations will be discussed in more detail in the next chapter on restorative justice.
- To respect internationally agreed guidelines on the fair procurement of biological resources and traditional knowledge, e.g. the Nagoya Protocol (Conde Gutiérrez 2018). Many of these guidelines are also supported by principles of justice in exchange and restorative justice.

A central aim of procedural justice in our case is for innovation processes and outcomes to fulfil the demands of the four discussed forms of justice. To facilitate such processes, we need an overall consensus on what people can expect from innovation systems, general guidelines on what constitutes a fair process, and specific principles for the major issues at stake, such as the protection of autochthonous varieties, the acquisition of traditional knowledge and the conservation of slowly renewable resources. Let us review these elements one by one.

8.3 Establishing Responsive Research Agendas

As mentioned before, agriculture is embedded in a dynamic social and natural environment to which it constantly has to adapt in order to maintain or increase yields. It is therefore essential for all agricultural innovation streams to embrace the fact that they have to be constantly responsive to new environmental and social needs and circumstances. History shows that societies have paid an enormous toll in human suffering and lives for their failure to adapt to changing environmental conditions (McMichael 2017). Here we need to keep in mind that it is the poor who pay the highest price for unsustainable food politics. Since social and environmental background conditions continuously change, and it is difficult to foresee beyond a few years what research can offer farmers and on what terms (van den Belt 2015), we need fair processes to address social justice questions as they emerge and become foreseeable. In other words, a changing natural and social environment demands a dynamic innovation assessment system to identify and respond to emerging challenges and opportunities.

The demand to continuously adapt innovation agendas to new social and environmental conditions has been referred to as making such systems "responsive". In a wider sense, such responsiveness has three major dimensions: (i) reducing risks and harms, (ii) addressing people's current and future needs, and (iii) exploring and developing ideas to improve well-being. In European policy discussion, the establishment of responsive research agendas is seen as one of the fundamental requirements of Responsible Research and Innovation (Stilgoe et al. 2013).

When it comes to assigning duties regarding the establishment of responsive research agendas, there are different ethical recommendations for private as opposed to public research. Utilitarians are generally less likely to draw distinctions between private and public research, as the capacity to make positive changes towards improving welfare is seen by them as central. In contrast, liberal approaches are far less demanding for private parties, as they will only be obliged to respect and not interfere with the state's obligation to secure food needs. Yet, irrespective of funding sources, it remains imperative that everyone adheres to a principle of non-harm. From a policy perspective, it is problematic that the way that non-harm is interpreted varies considerably (Holtug 2002). Few—mostly utilitarians—would maintain that people are harmed when opportunities to help are not fully exploited. A more generally accepted idea is to interpret a duty not to harm as making sure inventions can be safely used and do not directly cause harm to humans or other non-target organisms (so, for example, they are non-invasive for wildlife). Differences can also be found in how far the interests of future generations need to be taken into consideration. Some economists argue for future discounting, in the sense of calculating harms as less significant if they occur in the distant future, since people in the future are believed to have increased technical capacities to adapt to future harms. Others reject the idea of future discounting, especially when it involves the destruction (or deterioration) of irreplaceable assets (Goodin 1983) such as biodiversity. In general, perverse or misplaced incentives can lead to completely different results from what benevolently acting individuals would welcome. It is crucial for procedural justice that we set up institutions and inquiry processes that systematically assess the various incentives that innovators have in view of their effect on society, so that profit-maximizing entities maintain socially and ecologically acceptable practices while competing with each other.

When it comes to publicly funded research, it is imperative that agendas direct research to respond to present and future foreseeable needs. Here it is important to recall the various commitments that governments have agreed upon, which come with responsibilities towards their own citizens and duties to assist people in need abroad. To address such needs the International Covenant on Economic, Social and Cultural Rights (1966, art. 11.2.a) specifically calls for international cooperation (De Schutter 2011). Responsiveness should, however, go much further than minimizing risk and securing basic needs. As commentators on the human right to food have made clear, people have a right not only to food, but also to adequate food, which allows people to object to certain foodstuffs on non-nutritional grounds (UN Committee on Economic Social and Cultural Rights 1999). In terms of nutritional grounds, innovation needs to embrace the fact that humans do not only need calories, but are in need of a wide range of nutrients, whose absence leads to health deterioration and limitations in full development (Ickowitz et al. 2019). To be truthful to principles of self-determination, society should have a right not only to object to certain technologies, but also to have a say on choosing whole research directions, including which food future we want to have. For example, society could choose to adopt agroecology and improve the conviviality of human-made and natural ecosystems, or continue to improve conventional agriculture, reducing its negative effects and increasing yields in order to grow food in less space, and developing policies to ensure that the land that thereby becomes available is indeed given back to nature.

Innovators need to already consider in the design process the social mandate of making the fruits of research accessible by giving preference to research lines that could deliver cheaper or easily replicable options. In parallel, a critical assessment is needed to see to what extent different research incentives are conducive to the results expected by the public, and to fill in research gaps to facilitate a transition to more sustainable systems.

It is important under this form of justice for major decisions to be made democratically, after informing the public and establishing legal safety barriers to respect minorities and the foreseeable interests of future generations. To make sure research agendas are responsive to social needs, stakeholder involvement is usually seen as crucial, and this requires adequate communication and efforts to improve citizens' scientific literacy.

8.4 Communicating Research Programmes and Outcomes

To allow a meaningful debate with the wider public on innovation assessment, scientific and political literacy needs to be improved by facilitating public outreach and education campaigns. From the science and technology side, meaningful participation requires scientific projects to be sufficiently publicized, so the interested population can know what is going on and what the aims of the different projects are. This also benefits the advancement of science, as it facilitates cooperation, lets outsiders give feedback and report possible duplications or dead ends, and allows people with similar ideas to change or adapt their research plans. On the science governance side, this transparency allows the public to carry out parallel studies on risks, inform themselves through alternative channels and build technology support or opposition strategies after an informed independent assessment (Manca et al. 2017). Transparency is particularly important for regulatory science, as both the public and industry can gain knowledge of which technologies were approved or rejected and on what values such judgements were based (Fernández Pinto and Hicks 2019). From an economic perspective, communication allows technology developers to identify opposition and fears at an early stage, allowing them to respond to eventual worries by engaging in a public dialogue, or in cases of persisting opposition, abandoning the technology (Asveld et al. 2015; van den Belt 2015).

A failure to communicate research results is particularly blameworthy when experiments were performed with help from the community. Reporting back to participants facilitates transparency and improves accountability, although special care needs to be taken that the provided information is diffused responsibly, so as to avoid being misinterpreted or misrepresented (Taylor 2019). Beside these instrumental reasons, participants have a right to know for what purpose they have contributed their time, resources and effort.

There are cases in which the communication of research results poses a significant danger. In the case of the proliferation of information about the compound on which the antifoliant “Agent Orange” is based, Jacob and Walters (2005) assign substantial responsibilities to the inventor and even the publisher for making such dangerous information widely available. Unless such publication creates a massive backlash, both gain financial benefits by sharing information about the chemical compound. These authors are right to note that the publication itself is a substantial danger, as this information can be used for the development of warfare, despite the specific international prohibition on using hunger as a weapon of war by targeting food production systems. This information can also be used by terrorist organizations to cause harm. It should be noted, however, that a proper assessment and social response to such inventions requires some level of diffusion. This information is crucial for developing measures to address its toxicity. Even if information about the compound and its production is classified, protection measures still need to be developed in case the information is leaked or stolen.

Efforts to improve political literacy should not be neglected. When establishing research agendas, doing scientific work, diffusing science and engaging in public consultations, scientists and the wider citizenship need to be aware of the different types of injustice we have discussed and their demands, major international commitments towards social justice, specific regulations to avoid harm, and the main principles of good conduct within public deliberations. A failure to provide such political and scientific literacy will mean that many of the transmitted messages will fall on insufficiently receptive ears. The success of communication strategies depends on developing, in parallel, the audience’s skills as critical listeners.

Lastly, as we saw in the previous chapter on intergenerational justice, a transition towards long-term sustainability requires people—regardless of whether they engage in research activities—to acquire a basic understanding of collective action problems. Our population numbers demand that we reflect on how our actions affect, on an aggregate level, the lives of others and the environment.

8.5 Facilitating Participation

The most central feature of procedural justice is demanding sufficient participation in decision-making. In our case, this particularly concerns giving people a say in setting the direction of innovation agendas and on technology use. Procedural justice seeks meaningful participation, and credibility encourages us to make reasonable efforts to invite outsiders to participate (Ottinger 2013), but under the most common interpretations it does not go as far as contributive justice in demanding participation from people, nor in asking people to develop skills to be able to participate meaningfully. Here I defend a broader understanding of procedural justice that recognizes the extreme inequalities that exist among people, as well as the fact that people are not born with the necessary capabilities to promote justice and defend democratic institutions. This requires assisting people to develop a sufficiently high degree of scientific and political literacy, to empower them to protect their own interests and defend the rights of others (Atenas and Havemann 2019).

Participation has both intrinsic and instrumental value. The act of participating allows the use of skills, the ability to engage in dialogue, opportunities to socialize and network, and leads to increased knowledge. Moreover, it is valued as a sign of being taken seriously and recognized as a discussion partner. Innovators may also have an interest in participation, as it contributes to making their innovation known, provides room to clear misconceptions and address fears, allows people to identify future needs and unforeseen uses, and permits innovators to change research trajectories or abandon innovations that will meet excessive public opposition or involve considerable social and environmental risks. The wider public can play a role in shaping research agendas in a desirable direction and demand that certain risks and annoyances be addressed in a timely manner (Delgado 2010). Furthermore, common projects allow people to build a community in which experiences and knowledge are shared, and thereby constitute much more than an approach to acquiring knowledge.

The idea of understanding stakeholder involvement as an issue of justice has deep roots in the environmental justice movement (Schlosberg 2013). Yet it is not enough for stakeholders to be involved. In order for procedural justice to be meaningful, they should have a real opportunity to object to research trajectories and the release of certain innovations. This requires access to information, to be able to exercise informed consent (Shrader-Frechette 2002). Here the population groups who are directly affected by innovation should also be recognized as potential providers of local knowledge, as they might have experienced or observed phenomena that need to be considered in decision-making. It is crucial that discrimination is actively opposed and that institutions ensure that historically disadvantaged communities are able to exercise their voices. It is also important for innovation that involves many uncertainties and much incomplete knowledge—as is the case in the life sciences—that the public is informed of new relevant knowledge, and that they have the opportunity to retract their consent on the basis of this new information and new local knowledge (Vermunt and Steensma 2016), such as first-hand experience of detrimental effects. People should be able to revise their consent (and previous rejections) on the basis of new evidence. Cases where consent needs to be retracted are far from hypothetical in agriculture, as the prohibition of approved pesticides clearly illustrates (Kroma and Flora 2003; Cabrera 2017). To maintain incentives to participate in consultations, decisions taken from deliberations should generally lead to political change and not only have symbolic value.

It is important that outsiders are not only able to add content to the discussion and make certain objections, but that they also have a say on what is to be discussed and how the discussion is framed (Álvarez and Coolsaet 2020; Hicks 2017). People must have the opportunity to question and change how questions are asked, the labels that certain technologies receive and the direction that innovation should take. People should have the power to question labels and demand revisions. Labels may come with significant emotional baggage. For example, a large part of the debate on "biotechnology for the poor" implies that by making some modifications in policies and adapting technologies, one could solve the problems of the poor through biotechnology (Glover 2010). By framing this discussion with such labels, major changes such as opting for other types of innovation trajectory are often avoided, thereby concentrating the discussion on the optimization of innovation and its governance instead of democratically deciding on a specific research trajectory. The capacity to frame a discourse has been identified as a form of power, as it is something that is commonly held by those with stronger political and economic means, and lacked by those without such means (Hisano 2005). A similar phenomenon has been identified in how the debate is framed around conventional agriculture and agroecology. Here the question asked is how conventional agriculture can feed the world, and in relation to an alternative trajectory, the question is rephrased as what role agroecology can have in contributing to food production (Tittonell 2014). This bestows the lead role to conventional agriculture and assigns agroecology only a complementary role. Even global agricultural development indicators use questionable metrics that give conventional agriculture a major advantage, by focusing on the yield of particular staple crops per surface area, instead of calculating the overall farm productivity by considering all edible crops and fodder (Shiva and Pandey 2006). In a world with such high commercial interests in agriculture and a highly consolidated agrochemical and seed industry, it is difficult to regard the use of such advantageous metrics as pure coincidence. It must be noted that there are also cases where huge biotechnology corporations have experienced the costs of polemic labels coined by civil society organizations. For instance, when Monsanto acquired patents for genes with the potential for developing infertile seeds, opposing groups nicknamed these "terminator seeds", provoking major public outrage (Thompson 2009).

Education campaigns for both the general public and scientists are key to avoiding unnecessary polarized positions. To support informed decision-making, governments need to fill eventual knowledge gaps concerning environmental risks, revise information provided by the industry, and get involved in citizens' science education. Likewise, to avoid the underuse of safe innovations, fears and misbeliefs need to be addressed. This requires measures to provide impartial information where applicable, and efforts to detect and stop the propagation of fake news or information without adequate scientific backing. Furthermore, despite the fact that full participation is a mere ideal, due to the cognitive limits of full participation (Lafont 2003), there is still much more that governments and scientists could do to facilitate wider participation. Needlessly complex concepts can be simplified, versions with less formalizations and better graphical descriptions can be made available, and specialists can be trained and hired to improve the public understanding of science (Popa et al. 2020).

To maintain active participation, people engaged in a dialogue need to respect each other and be open to changing their opinions and listening to new information (Kuyper 2015). To build towards change, an open dialogue requires each participant to be capable and willing to critically analyse and reflect upon radically different points of view and revise their convictions and actions accordingly (Lauer 2017; Keulartz et al. 2004). This respect and openness needs to be offered on a reciprocal basis. Participants should also value different perspectives for their epistemic content, as much knowledge is local, and drawing from a dissimilar pool of life experiences can lead to a larger range of problem and solution identification possibilities (Solum 2004). Dissenting opinions and standpoints are particularly valuable when the majority becomes discontent about the current state of affairs (Anderson 2009), as building on these ideas is more advantageous than starting from scratch. To maintain the flow of diverging ideas it is crucial that every individual and group can introduce them in the relevant forums and announce their concerns. To maintain this diversity, we need to maintain parallel innovation streams that place special value on examining microhabitats and niche conditions (Lebel and McLean 2018). Taking into account the extreme inequality that exists between people both within and between countries, it will take much work to encourage people from all ways of life to participate and be heard. Here we need to recall the different forms of discrimination that we discussed in the chapter on contributive justice. Then there is also the possible factor of shame. Amartya Sen identified the importance of the capability of being in public without shame (Sen 2009), which for many scientific and political events would include being dressed according to professional or local customs: an urge that might be stronger among those from a poorer social background. This will require some redistributive measures to facilitate participation from people with less material means. In cases where a dialogue is pursued via representatives, they should reflect authentic values and allow people to identify with them and their voice.

As intellectual property rights expand over scientific areas in a world of increasing inequality, attention also needs to be paid to ensuring that people from all over the world can access publications and research materials legally. Although some people consider the whole scientific publication system as unfair, and have little moral scruples about illegal downloading and content sharing, it is highly likely that those people would prefer a scenario where they could access these contents legally, as only a minority of people pirate or contraband for the mere thrill of it. People do feel distress when they have to live under conditions where they cannot avoid breaking laws on a regular basis to participate in communal life, and it is an injustice to not improve their situation (Wolff and de-Shalit 2007).

There is an ongoing debate about whether or not participation should be compulsory. With respect to democracy, some countries have made voting (or at least showing up at polling stations) compulsory, as is the case with Argentina, Australia and Belgium, among many others. There are many arguments that back mandatory voting for the governance of common interests. For instance, democratic governance can be understood as a public good to which all should contribute—failing to do so would be to free-ride on others' efforts. Moreover, one could argue that democratic involvement in decision-making contributes to decision ownership, which has the effect that people would back commonly agreed policies with their advantages and disadvantages. Another argument highlights the epistemic importance of large-scale voting. A massive number of voters is likelier, on an aggregate level, to identify the most optimal policies, compared to a panel of experts (Reiss 2019). Regardless of these benefits, most countries do not have mandatory voting polices, or else fail to enforce them. The liberal principle of avoiding any form of compulsion takes precedence in this case, despite the fact that many of these countries support mandatory jury services or conscription. Furthermore, it has been argued that gathering the necessary information and analysing the options in order to vote responsibly takes considerable effort (Brennan 2011). To check that people do indeed take the necessary steps to vote responsibly would require an extensive invasion of privacy, and obliging uninformed voters to go to the ballots could be harmful for society. Therefore, the preferable option would be to leave it to each person's discretion when to participate in democratic decision-making or take other types of political action. Independently of what one's political stance is on this question, it still needs to be stressed that there is a difference between not making voting mandatory and being indifferent about low levels of participation. The latter can be a symptom of apathy or a complete lack of confidence in public institutions. Procedural justice demands that society encourages participation.

Lastly, in order for participation not to lead to stagnation, it is crucial for political literacy to be improved. People need to learn about the proper channels to file complaints, offer advice, seek information and call for political action (de-Shalit 2004). While deliberation should aim at achieving consensus, there will be cases where this will not be fully possible. Yet to participate in political forums, it is important for people to make compromises on decisions achieved by majority rule, and that they only raise further objections when these are backed by major concerns about basic rights, fundamental values and intergenerational justice. Fighting endlessly in courts to avoid or postpone inconveniences stalls innovative processes and larger projects. Consultations are only meaningful when people are willing to accept the different proposed alternatives. Lastly, people also need to be aware that procedural justice is imperfect. Justice is only one of the major interests that people have, and increasing the accuracy of the outcomes of procedures makes them more and more complicated to master, and augments the price of establishing them (Solum 2004).

8.6 Risk Management and Gatekeeping

A key element of responsible technology introduction, use and disposal is to have adequate policies regarding risk management and communication. This requires not only mechanisms for responding to risks, but also procedures to prevent population groups or ecosystems becoming endangered. This requires risk identification, guiding principles for handling possible risks, and constitutional safeguards to avoid any outcome where the interests of minorities, minors and future generations are ignored.

Risk assessment requires the identification of risks. This is a major challenge, as it prerequires a vast amount of scientific knowledge that we only partially possess, while some parties have significant financial incentives to remain ignorant about some of these possible risks, or to keep possible risks secret (Fernández Pinto 2019). We need studies and trials for the assessment of risks. However, some of these studies in themselves already involve a risk. An extensive risk assessment requires that we do not only test technologies in a closed laboratory, but rather carry out site-specific tests to identify any possible unforeseen side effect that may emerge as the technology (i.e. the agrochemical or modified variety) interacts with other organisms or substances. In addition, there is the problem of self-replication (Lim 2013). Living organisms tend to multiply, and some newly introduced organisms mix with native organisms—raising questions of responsibility that are difficult to settle (Robaey 2016a). The failure to use a technology may also pose a problem, particularly with respect to technologies that are strongly needed to overcome urgent problems, such as climate change adaptation and improving food security (Thompson 2018). In a world as technologically interconnected as ours, leaving things as they are also poses a risk.

Nowadays we are increasingly in need of constitutionally based rights or special enforceable conventions to limit the harmful or socially undesirable effects of innovation (as mentioned in Chap. 7). In the European context, we can observe a strong demand to adhere to a common principle regarding risk management: the precautionary principle. One could even demand the inclusion of this principle in decision-making as a matter of procedural justice (Schomberg 1997). Nonetheless, from a procedural justice perspective, it is crucial that there is transparency and consistency in the justification for excluding technologies on the basis of a risk assessment. While a community may reserve the right to object to a technology for cultural and social reasons, these reasons have to be labelled as such, and not be hidden behind insufficiently supported or false scientific arguments that claim that the technology poses a greater risk than other socially preferred technologies (cf. Fernández Pinto and Hicks 2019; de Melo-Martín and Intemann 2012). In addition, when precautionary measures are taken to avoid social and environmental harms, special care needs to be taken that measures for being cautious about a certain risk do not blind regulators to other larger risks, or underestimate the disadvantages of not using the technology at all (Sandin 2004). The decision over which principle or set of principles to adopt for technology assessment would have to be taken democratically, as people have different levels of risk aversion. Reaching a consensus on what is an acceptable level of risk for humans and the environment is particularly difficult, as it necessarily involves value judgements over which large population segments will disagree (Bovenkerk 2012). However, the times are gone when it was enough for each country to worry about environmental policy within its own borders. An international dialogue is strongly needed to align the different strategies used to minimize risks, as many agricultural policies have implications beyond borders, as is the case with the introduction of new species, the provision of ecosystem services, and climate change (IPBES 2019).

In addition, we need safeguards to ensure that majority or powerful interest groups do not overrun weaker parties or future generations (Villarroel 2013). One option is to give underrepresented and unrepresented groups a special vote weighing or voice, either by making their vote count for more, or by instituting special members of parliament to vote on behalf of those who are unable to vote (cf. Gosseries 2008; Zakaras 2016). People who are highly vulnerable to climate change or agrochemical runoffs, such as indigenous people and other minorities, would need such protection to prevent their interests being overrun by inconsiderate majorities (Schlosberg and Carruthers 2010). Future generations are also in need of such protection, so that they can secure their right to food. It has become clear that the strong demand for cheap food—especially cheap meat—is having disastrous consequences for the environment and animal welfare (Singer 1975/2009; Patel and Moore 2017). Innovators need to commit to a "minimum irreversible harm" principle, in order to reduce unintended negative consequences that could jeopardize the welfare of future generations (Wallack 2006). This would require innovators and governments to carefully check that innovations do not use resources beyond regenerative capacities, and also to provide adequate compensation for the use of exhaustible resources.

8.7 On the Social Value of Knowledge

Particularly in relation to biomedical research involving human research subjects, it has been claimed that research should lead to social value (Holzer 2017). The reasoning behind this claim is that researchers who intervene in members of a community inflict certain risks, and that as a form of compensation, they or their community should receive some benefits. Failure to provide benefits, or inflictions of harms, would be seen as exploitation or a risk factor by those directly affected and by other members of the community, in extreme cases even leading to mistrust or scepticism towards everything related to science. In addition, failing to attend to pressing public needs while engaging in scientific projects with little or no social value reduces public support for science, which may ultimately also negatively affect the ability to develop socially useful research (Intemann and de Melo-Martín 2014).

There is also the argument that a community becomes complicit by facilitating a research experiment. In this sense, they gain an interest in ensuring that the research they played a role in is used only for beneficial purposes. For instance, biosafety tests should lead to reducing harm to people and the environment. When incidental findings show that the tested substances have a high level of toxicity, measures need to be taken to ensure that these results are not weaponized or used for other harmful means. For example, we may assume that smallholders who have stewarded and developed a particularly hot chilli pepper would most likely be enormously upset if they were to find out that their variety has been used to develop a stronger pepper spray used to fight peasant demonstrations, among others. It must be assumed (unless explicitly specified otherwise) that people would not have consented to participate in research that is destined for harmful uses.

Nonetheless, insisting on social value should not be understood too narrowly—that is, in terms of immediate benefits—and should allow enough room for scientific curiosity and fundamental research. Policies that urge scientists to produce a rapid "valorisation" of research make it difficult to depart from certain path dependencies or offer radically new approaches, as it is often impossible to foresee such outcomes when one has to justify extensively each step taken and ask each time for separate funding.

8.8 Increasing Access to the Objects of Innovation

As discussed in earlier chapters, human rights declarations establish that people have a right to benefit from scientific advancement (Marks 2011; Plomer 2012). Access to innovation is in many cases crucial for being able to benefit from scientific advancement, which creates obligations to consider issues of accessibility already in the innovation proposal and design phase. The choice of materials and even the direction of whole research lines have an effect on who will be able to access the fruits of innovation. To meet this demand, innovators will need to continuously work towards making their innovation more accessible and be particularly open to ideas from outsiders on how to reduce costs and improve distribution channels. Technology users have been particularly innovative in improving the technologies they use and finding new uses (Torrance and von Hippel 2015). To be responsive to public needs, innovation needs to acknowledge resource constraints and incorporate already in the design phase ideas that will allow sufficient diffusion (Levidow and Papaioannou 2018). Such strategies go hand in hand with efforts to achieve justice in exchange for innovation acquisition.

There is also a widespread demand for making knowledge accessible. Major examples are the access to knowledge movement (Kapczynski 2008) and initiatives to protect the public domain (Boyle 2008). Here access to knowledge is valued for both its instrumental and its non-instrumental value, that is, for empowerment and for mere curiosity. Addressing this demand will also require making information available in early design phases and not only at the end stage, when the products are being put on the market. When research has received funding from taxpayers, we can observe an even stronger demand for access, since the public claims to have a right to see what was done with their money. In addition, since all taxpayers contribute to the funding of research and the maintenance of infrastructure through taxes and often also work, there is a justice argument for minimizing access hurdles.

As the role of the private sector increases, we may wonder whether there is a social obligation to strengthen a public sector that offers innovations on a public or common good basis, in order to reduce the negative effects of inequality and competitive disadvantage (Bragdon 2016). Such innovations could allow the continuous livelihood of smallholders who still produce a substantial amount of food, have lower food waste, and provide valuable ecosystem services (cf. Ricciardi et al. 2018; van der Ploeg 2010).

8.9 Sustainable Research Praxis I: Handling Scientific Misconduct

Despite some considerable occasional drawbacks, over recent decades we have seen a major movement towards making scientific practices fairer and reducing the likelihood of abuses in scientific work environments. The fact that science was done for so many centuries in highly secluded environments—and under the usually justified belief that the state and church should not interfere, since they may hinder the pursuit of knowledge, and often also under highly hierarchical structures—has delayed attempts to establish good scientific practice beyond self-regulation. Nowadays, research ethics is a continuously evolving research field that identifies injustices and offers recommendations as new challenges continuously emerge (Koepsell 2016; Anderson 2005). Codes of good scientific practice or good conduct are becoming fairly common in all areas of research. Within the wider concept of research integrity we can distinguish between epistemic concerns—i.e. epistemic integrity, discussing issues such as data fabrication and falsification—and ethical problems—different forms of plagiarism, authorship disputes, failure to obtain informed consent, sexual harassment, and multiple forms of discrimination (De Winter and Kosolosky 2013).

Outside biomedical research, for a long time research integrity concentrated on the prohibition of the falsification and fabrication of results and plagiarism. Respecting the moral interests of authors (see Chap. 2) demands that all intellectual contributions are given proper credit. Nowadays, this also requires considerable efforts to discover misattributed authorship, and especially to acknowledge the often misattributed, neglected or erased contributions of minorities and female scholars. Addressing moral interests has both ethical and epistemic functions. One is that proper recognition is a desirable good that also amounts to an incentive to carry out research. Another is that by linking a scientific contribution to a name that someone took considerable effort to build, we encourage responsibility regarding scientific accuracy and validity (cf. Rolin 2015). It thereby contributes to the major social and scientific effort to reduce the falsification and fabrication of data.

We are increasingly faced with the demand to include prior informed consent among the procedures required by good scientific practice. This principle, which gained wide acceptance in post-Word War II medical ethics, has taken considerable time to be adopted in other fields. Nowadays it is being proposed in areas as diverse as environmental studies and rural sociology. People who collaborate with researchers, sharing observations and denouncing wrongful activities of industry and governments, are often in real danger, risking torture or even their lives. A related issue concerns deceit. Here intention is the distinguishing factor that condemns deceiving as actual malpractice. There are instances where users and other researchers have been deceived by the inclusion of too much or too little information, by downplaying the role of disadvantageous information, by omissions or selective publication, or by hiding the fact that users are participating in an experiment.

An element that continues to be of major concern comprises conflicts of interests. Many jurisdictions have regulations to disclose financial conflicts of interests, but there is no international consensus on how to handle other types of conflict of interest (Resnik 2009). Here we may consider specifying potential biases, for example due to collegial, mentor-tutor, institutional or personal relationships that may affect the authors' judgements. Such biases may have major consequences for epistemic integrity: depriving society and the scientific community of distinctive approaches and cognitive diversity, making it less likely that established methods and procedures are adequately examined, and hindering a fair competition of ideas.

The scope of what good scientific practice consists in is expanding considerably. Particularly when research involves experimenting with sentient beings, we have already seen a tendency during various decades to prohibit and reduce research that causes disproportional stress or pain, both physical and emotional (Singer 1975/2009). In relation to global justice, the principle that participation in research should also be of benefit to the research subjects and their community is also starting to be accepted outside medicine, as the access and benefit sharing principle illustrates (Schroeder and Lasén-Díaz 2006). Slowly, an ethical mandate calling for the diversification of literature to acknowledge authors from historically disadvantaged groups is becoming an issue (Muldoon 2019).

Addressing scientific misconduct will require making major structural changes. Hierarchical structures and weak labour protection laws for younger scientists and temporary research associates create an environment that makes scientific misconduct, including even asocial and despotic behaviour, easy to continue unpunished, especially if carried out by senior male staff. In a recent survey, 21% of Ph.D. researchers reported that they had personally experienced discrimination or harassment, and the same proportion of respondents claimed to have suffered from bullying (Woolston 2019). Such work structures leave whole population groups at the risk of exploitation or abuse. Efforts must also be made to eradicate nepotism. Educational campaigns and sanctions must be effective enough to counteract deeply entrenched private interests and communal virtues, such as loyalty and camaraderie, whose exercise counters efforts to improve public welfare by hiring the best qualified applicants (Metz 2009; Walzer 1993). This also requires addressing the biases that researchers and members of hiring and funding committees may have for or against particular groups of people, including their own (Wray 2007). Such biases not only lead to injustices by disadvantaging researchers, but also bring epistemic poverty by homogenizing the research community and research approaches. People interested in maintaining or improving the reputation of such structures must establish measures to avoid injustices. In some cases, this will also require them to relinquish some of their current power to avoid further abuses.

Distrust in science has negative effects for social justice and sustainability. Malpractice erodes public trust in science: something that may take years or decades to rebuild. This has a huge effect on public health and the environment when citizens do not take scientific advice seriously, fail to approach specialists with their concerns, and fail to play their part in limiting the propagation of pathogens.

8.10 Sustainable Research Praxis II: Fair Procurement of Genetic Resources and Traditional Knowledge

Recognition of the extremely unequal power relations between indigenous or traditional communities and multinational corporations—in terms of financial resources, technical expertise and political influence—has created a strong demand for procedures that stop or at least reduce the worst forms of exploitation (Marion Suiseeya

2014). As a matter of justice, there is a strong demand for establishing legal parity between traditional knowledge systems and conventional scientific systems (Desai 2007). In the spirit of the principles of procedural justices and justice in exchange, we can observe global treaties that aim to lead to fair access and benefit sharing. Such treaties welcome the outcome that genetic resources and traditional knowledge are acquired by others, on the condition that prior, free and informed consent was obtained, and that the providers and stewards of such resources are given proper acknowledgment and a fair share of the benefits developed from their contributions (De Jonge 2011; Schroeder and Pisupati 2010).

The Nagoya Protocol is currently the most extensive treaty concerning genetic resources and related traditional knowledge. Principles of access and benefit sharing are embraced, as a matter of justice—in terms of fair rewards for conserving, identifying and developing a variety and traditional knowledge—to recognize communities for such work, and as an instrument to encourage further dedication to such efforts. This principle seeks to depart from past practices—still prevalent—that treat indigenous genetic resources and traditional knowledge as common property, or even as something that is unowned and can be freely used as pleased (Timmermann and Robaey 2016; Millaleo-Hernández 2019).

Efforts to secure fair rewards should not, however, lead to a reduction in scientific collaborations and the exchange of genetic resources between the Global South and Global North, as this is likely to negatively affect the resilience of food systems and future food security (Deplazes-Zemp et al. 2018; Prathapan and Rajan 2011). Here it is particularly important to appeal to the obligations of countries, particularly in the Global South, not to mismanage resources and to create favourable conditions for international cooperation (Barugahare and Lie 2014), in order to allow mutually beneficial exchanges.

8.11 Protection for Whistle-Blowers

Due to the high financial stakes, and the enormous social importance of certain agricultural innovations, there is strong pressure to hide malpractice and even criminal conduct. Universities have been accused of hiding cases of sexual misconduct and plagiarism by senior faculty members, at a serious cost for the victims, as well as for staff who want to work in a respectable institution. Cases of data fabrication and falsification are sometimes hidden for years, and some cases of bribery are given a low profile. Environmental activists have paid a high price for their courage in denouncing polluters, poaching activities and illegal logging of endangered forests—sometimes even with their lives. To protect conscientious denunciation, procedural justice needs to provide safeguards for informants (Dekker and Breakey 2016). Such protections need to carefully balance private interests in privacy and business strategies that rely on trade secrecy with the strong public interest in self-determination, safety, protecting the law and fighting the corruption of public officials and institutions (Santoro and Kumar 2018). Here we need careful deliberations on what constitutes the public interest, in order to defend people who deviate from any contractual obligation to keep information confidential, to avoid creating a loophole for people to claim to have breached their contracts for the public interest, when they were mostly guided by private interests (Boot 2020).

8.12 Tackling Excessive Influence

A major challenge for procedural justice is to keep processes fair under extreme inequality. We are currently witnessing cases of the most extreme concentration of capital in history, leading to enormous inequalities in financial resources and political power (Piketty 2014). At the same time, we can also observe a massive consolidation of the food and agriculture sector, often leading even to complete control of food supply chains, from agricultural inputs to supermarkets (Ziegler 2011; De Schutter 2017). Many multinational corporations already have more financial resources than some small countries. This enormous concentration of power, through purchasing capacity and the ability to retain and hire specialists, allows specific companies and individuals to distort democratic decision-making procedures, supress information that is detrimental to them, flood scientific publication outlets and media with information that benefits them and casts doubt on data that affects them negatively, and even buy political influence by bribing public officials or making them dependent upon their donations.

A strategy that is often used is to fabricate controversies that have nothing to do with legitimate scientific dissent in order to delay regulation and gain sympathy from the wider public (Hansson 2018). Sometimes their influence can be hard to perceive. For instance, a multinational corporation can deliberatively make available massive amounts of money to study a particular research area that does not conflict with their interests, in order to divert research attention from areas that could jeopardize the sales revenues of their products (Fernández Pinto 2019).

To counter such detrimental influence, we need clarity on the diverse sets of practices that constitute corruption beyond the obvious cases of bribery, in order to learn how to tackle them (Ceva and Ferretti 2018). The ways in which corporations build socially and environmentally detrimental influence and obstruction have multiple facets. For instance, we also need specialists to prevent corporations and key actors strategically dividing civic society organizations in order to destroy their influence through the old *divide et impera* tactic (cf. Galtung 1971), especially by creating differences and conflicts between biodiverse rich countries and farmers' organizations, to gain the advantage of being able to exert stronger pressure on divided groups. Lobbies are already a huge threat to environmental regulation and democratic decision-making. It is crucial that governments maintain or regain their power to control the different corporations, institutions and organizations who control parts of the food system, and make them accountable to democratic decision-making (Anderson et al. 2019). To regain democratic control over food production, citizens and farmers also need to be empowered to exert pressure, as individuals and communities, to counteract the domination of major food corporations (Bornemann and Weiland 2019). Members of under-represented communities need to learn how to identify and object to structures and procedures that divide smaller interest groups and make it impossible for small delegations to gain sufficient oversight over negotiations and debates (cf. Bragdon 2016; Marion Suiseeya 2014). Moreover, civil society and farmers' organizations need to learn that not all forms of cooperation with groups one opposes are necessarily a form of complicity. It is important that even mutually opposing groups explore win–win strategies after carefully assessing the benefits and harms of cooperating (Jauernig et al. 2019), and contrasting this outcome with the counterfactual scenario of inaction and maintaining the status quo. For example, while input-substituting organic farmers and agroecologists have major differences of opinion concerning farm system design, this disagreement should not hinder them from working together to protect pollinators.

8.13 From Procedural Justice to Restorative Justice: Allowing Coexistence and Encouraging Mutually Beneficial Interactions

Our current agricultural systems and environmental crisis oblige us to reconsider how we nowadays produce food (McIntyre et al. 2009). We need to establish procedures that allow the coexistence of different farming systems, which requires that people sometimes refrain from using certain types of innovation, or that users of certain innovations compensate for the ecosystem services that they rely on others to produce. For example, the developers of pesticides assume that there are other farmers who abstain from using their products and provide a safe harbour for species crucial to human survival, such as pollinators and birds. This service needs to be compensated and its value recognized.

A crucial problem is that many agricultural innovations ignore boundaries. Some have an effect beyond farm boundaries, which has an impact on other farmers, forcing them to adapt their production methods (Tittonell 2016). Other technologies replicate themselves, such as genetically modified seeds, affecting neighbours when these crops end up growing in their fields, which may lead to a loss of sales opportunities (Robaey 2016b).

Improving agricultural sustainability is also subject to international treaties. Agriculture is responsible for a major part of the emitted greenhouse gasses, especially through transportation, the maintenance of ruminants and the production of fertilizers (IPBES 2019). Countries and regions who have committed to reducing their carbon footprint need to pay special attention to reducing their emissions from the food sector.

Although there are still considerable disputes over how we are to respond to the current crisis, and different farming systems are still under consideration, it is increasingly important that we direct our research agendas to stimulate innovation that allows and encourages mutually beneficial interactions. Ecological farming systems approaches in particular view a farm as embedded in a wider (agro)ecosystem (Francis et al. 2003). Here we need to consider closing nutrient cycles in agroecological farms and reintegrating organic material that was moved to the cities by composting and improving urban agriculture, therefore demanding innovations from both the natural and social sciences. This will also require the study and development of better associations between the different crop varieties in use, to allow a more beneficial interaction.

It would take a major effort to align agricultural innovation with the demands of social and environmental justice, and here there is a limit to how much we can achieve by establishing fair processes. Procedural justice is not failsafe: more inclusive deliberative processes are no guarantee for fairer outcomes (Blok and Lemmens 2015). The inclusion of a wider range and diversity of participants merely improves the chances that problems are identified and confronted at an earlier stage (de Graeff et al. 2018). Moreover, by including more people in decision-making, one improves decision ownership: the decision taken is no longer something alien, made by others. When people have taken part in a transparent deliberative process and had proper access to information about the dangers and advantages of technologies, they are likelier to accept responsibility for their decisions, and are more open to cooperating to work towards a solution. If technologies do not deliver what they were expected to, or cause unforeseen problems, the mechanisms that allowed inclusive decision-making for innovations can be used to work towards solutions in forums seeking restorative justice. Securing the different processes involved in procedural justice (Table 8.1) will make it much easier to secure any eventual demands of restorative justice, as the social institutions of the former can be put into use to address the conflictive issues

Table 8.1 Examples of processes for bringing about social justice

	Forms of justice	Examples
Processes promoting social justice	Justice in exchange	– Informed consent – Access and benefit sharing agreements
	Distributive justice	– Reorienting public research towards the needs of the poor – Diversification of incentive schemes
	Contributive justice	– Licenses that allow wider participation (e.g. creative commons) – Equality of opportunity policies
	Intergenerational justice	– Provision of public goods as compensation – Conservation measures – Establishment of ecological refuge zones

of the latter. Moreover, any failure to meet the demands of procedural justice will likely to have to be addressed later on in restorative justice processes, especially after technologies have caused major harms.

Procedural and restorative justice can be seen as two parts of a mechanism that seeks to implement the four aforementioned forms of justice: justice in exchange, distributive justice, contributive justice, and intergenerational justice. The former two are the means to achieve the latter four's demands. This, however, does not mean that they do not also have significant intrinsic value: people have an interest in the existence of such mechanisms and participate in them for their own sake, not just for the goals they intend to achieve. This is particularly important to keep in mind when assessing the inclusiveness of such instruments.

References

Álvarez, Lina, and Brendan Coolsaet. 2020. Decolonizing environmental justice studies: A Latin American perspective. *Capitalism Nature Socialism* 31 (2): 50–69. https://doi.org/10.1080/10455752.2018.1558272.

Anderson, Colin Ray, Janneke Bruil, Michael Jahi Chappell, Csilla Kiss, and Michel Patrick Pimbert. 2019. From transition to domains of transformation: Getting to sustainable and just food systems through agroecology. *Sustainability* 11 (19): 5272.

Anderson, Elizabeth. 2009. Democracy: Instrumental versus non-instrumental value. In *Contemporary debates in political philosophy*, ed. Thomas Christiano and John Christman, 213–27. Malden and Oxford: Wiley-Blackwell.

Anderson, Trevor R. 2005. The effective practice of agricultural science. In *Ethics in agriculture—An African perspective*, ed. Alvin Van Niekerk, 143–163. Dordrecht: Springer.

Asveld, Lotte, Jurgen Ganzevles, and Patricia Osseweijer. 2015. Trustworthiness and responsible research and innovation: The case of the bio-economy. *Journal of Agricultural and Environmental Ethics* 28 (3): 571–588.

Atenas, Javiera, and Leo Havemann. 2019. Open data sectors and communities: Education. In *The state of open data: Histories and horizons*, ed. T. Davies, S. Walker, M. Rubinstein and F. Perini, 91–102. Cape Town and Ottawa: African Minds and International Development Research Centre.

Barugahare, John, and Reidar K. Lie. 2014. Obligations of poor countries in ensuring global justice: The case of Uganda. *Etikk i praksis-Nordic Journal of Applied Ethics* 8 (2): 82–96.

Blok, Vincent, and Pieter Lemmens. 2015. The emerging concept of responsible innovation. Three reasons why it is questionable and calls for a radical transformation of the concept of innovation. In *Responsible Innovation 2: Concepts, Approaches, and Applications*, ed. Bert-Jaap Koops, Ilse Oosterlaken, Henny Romijn, Tsjalling Swierstra, and Jeroen van den Hoven, 19–35. Cham: Springer.

Boot, Eric R. 2019. The feasibility of a public interest defense for whistleblowing. *Law and Philosophy* 39: 1–34.

Bornemann, Basil, and Sabine Weiland. 2019. Empowering people—Democratising the food system? Exploring the democratic potential of food-related empowerment forms. *Politics and Governance* 7 (4): 105–118.

Bovenkerk, Bernice. 2012. *The biotechnology debate: Democracy in the face of intractable disagreement*. Dordrecht: Springer.

Boyle, James. 2008. *The public domain. Enclosing the commons of the mind*. New Haven and London: Yale University Press. https://thepublicdomain.org.

Bragdon, Susan H. 2016. Reinvigorating the public sector: The case of food security, small-scale farmers, trade and intellectual property rules. *Development* 59 (3–4): 280–291.
Brennan, Jason. 2011. The right to a competent electorate. *The Philosophical Quarterly* 61 (245): 700–724. https://doi.org/10.1111/j.1467-9213.2011.699.x.
Cabrera, Laura Y. 2017. Pesticides: A case domain for environmental neuroethics. *Cambridge Quarterly of Healthcare Ethics* 26 (4): 602–615.
Ceva, Emanuela, and Maria Paola Ferretti. 2018. The ethics of anti-corruption policies. In *The Routledge handbook of ethics and public policy*, ed. Annabelle Lever and Andrei Poama, 255–266. Oxon and New York: Routledge.
Conde Gutiérrez, Carlos. 2018. Una aproximación a la propiedad intelectual y el acceso a recursos genéticos desde la perspectiva del contrato social y la justicia global. In *Propiedad intelectual. Fundamento y crítica*, ed. Martín Hevía and Facundo M. Rojo, 145–198. Bogotá: Universidad del Externado.
de Graeff, Nienke, Léon E. Dijkman, Karin R. Jongsma, and Annelien L. Bredenoord. 2018. Fair governance of biotechnology: Patents, private governance, and procedural justice. *The American Journal of Bioethics* 18 (12): 57–59.
De Jonge, Bram. 2011. What is fair and equitable benefit-sharing? *Journal of Agricultural and Environmental Ethics* 24 (2): 127–146.
de Melo-Martín, Inmaculada, and Kristen Intemann. 2012. Interpreting evidence: Why values can matter as much as science. *Perspectives in Biology and Medicine* 55 (1): 59.
De Schutter, Olivier. 2011. The right of everyone to enjoy the benefits from scientific progress and the right to food: From conflict to complementarity. *Human Rights Quarterly* 33: 304–350.
De Schutter, Olivier. 2017. The political economy of food systems reform. *European Review of Agricultural Economics* 44 (4): 705–731.
De Winter, Jan, and Laszlo Kosolosky. 2013. The epistemic integrity of scientific research. *Science and Engineering Ethics* 19 (3): 757–774.
de-Shalit, Avner. 2004. Political philosophy and empowering citizens. *Political Studies* 52 (4): 802–818
Dekker, Sidney W.A., and Hugh Breakey. 2016. 'Just culture:' Improving safety by achieving substantive, procedural and restorative justice. *Safety Science* 85: 187–193.
Delgado, Ana. 2010. ¿Democratizar la Ciencia? Diálogo, reflexividad y apertura. *Revista Iberoamericana de Ciencia, Tecnología y Sociedad-CTS* 5: 15.
Deplazes-Zemp, Anna, Samuel Abiven, Peter Schaber, Michael Schaepman, Gabriela Schaepman-Strub, Bernhard Schmid, Kentaro K. Shimizu, and Florian Altermatt. 2018. The Nagoya Protocol could backfire on the global South. *Nature Ecology & Evolution* 2 (6): 917.
Desai, Pranav N. 2007. Traditional knowledge and intellectual property protection: Past and future. *Science and Public Policy* 34 (3): 185–197.
Doorn, Neelke. 2010. A procedural approach to distributing responsibilities in R&D networks. *Poiesis & Praxis* 7 (3): 169–188.
Fernández Pinto, Manuela. 2019. Scientific ignorance. *Theoria: An International Journal for Theory, History and Foundations of Science* 34 (2): 195–211.
Fernández Pinto, Manuela, and Daniel J. Hicks. 2019. Legitimizing values in regulatory science. *Environmental Health Perspectives* 127 (3): 035001.
Francis, Charles, G. Lieblein, S. Gliessman, T.A. Breland, N. Creamer, R. Harwood, L. Salomonsson, J. Helenius, D. Rickerl, and R. Salvador. 2003. Agroecology: The ecology of food systems. *Journal of Sustainable Agriculture* 22 (3): 99–118.
Galtung, Johan. 1971. A structural theory of imperialism. *Journal of Peace Research* 81–117.
Gliessman, Stephen R. 2007. *Agroecology: The ecology of sustainable food systems*. Boca Raton: CRC Press.
Glover, Dominic. 2010. The corporate shaping of GM crops as a technology for the poor. *The Journal of Peasant Studies* 37 (1): 67–90.
Goodin, Robert E. 1983. The ethics of destroying irreplaceable assets. *International Journal of Environmental Studies* 21: 55–66.

Gosseries, Axel. 2008. On future generations' future rights. *Journal of Political Philosophy* 16 (4): 446–474.

Hansson, Sven Ove. 2018. Risk, science and policy: A treacherous triangle. *Ethical Perspectives* 25 (3): 391–418.

Hicks, Daniel J. 2017. Genetically modified crops, inclusion, and democracy. *Perspectives on Science* 25 (4): 488–520.

Hisano, Shuji. 2005. A critical observation on the mainstream discourse of biotechnology for the poor. *Tailoring Biotechnologies* 1 (2): 81–106.

Holtug, Nils. 2002. The harm principle. *Ethical Theory and Moral Practice* 5 (4): 357–389.

Holzer, Felicitas S. 2017. Defending the social value of knowledge as a safeguard for public trust. *Bioethics* 31 (7): 559–567.

Ickowitz, A, Bronwen Powell, D Rowland, A Jones, and TCH Sunderland. 2019. Agricultural intensification, dietary diversity, and markets in the global food security narrative. *Global Food Security* 20.

Intemann, Kristen, and Inmaculada de Melo-Martín. 2014. Addressing problems in profit-driven research: How can feminist conceptions of objectivity help? *European Journal for Philosophy of Science* 4 (2): 135–151.

IPBES. 2019. *Summary for policymakers of the global assessment report on biodiversity and ecosystem services of the intergovernmental science-policy platform on biodiversity and ecosystem services*. Bonn: IPBES Secretariat.

Jacob, Claus, and Adam Walters. 2005. Risk and responsibility in chemical research: The case of agent orange. *HYLE–International Journal for Philosophy of Chemistry* 11 (2): 147–166.

Jauernig, Johanna, Ingo Pies, Paul B. Thompson, and Vladislav Valentinov. 2019. Theorizing agriculture-society tensions: An ordonomic approach to the agrarian vision. In *Sustainable governance and management of food systems: Ethical perspectives*, ed. Eija Vinnari and Markus Vinnari, 634–657. Wageningen: Wageningen Academic Publishers.

Kapczynski, Amy. 2008. The access to knowledge mobilization and the new politics of intellectual property. *Yale Law Journal* 117 (5): 804–885.

Keulartz, J., M. Schermer, M. Korthals, and T. Swierstra. 2004. Ethics in technological culture: A programmatic proposal for a pragmatist approach. *Science Technology & Human Values* 29 (1): 3–29. https://doi.org/10.1177/0162243903259188.

Koepsell, David. 2016. *Scientific integrity and research ethics: An approach from the ethos of science*. Cham: Springer.

Kroma, Margaret M., and Cornelia Butler Flora. 2003. Greening pesticides: A historical analysis of the social construction of farm chemical advertisements. *Agriculture and Human Values* 20 (1): 21–35.

Kuyper, Jonathan. 2015. Deliberative capacity in the intellectual property rights regime complex. *Critical Policy Studies* 9 (3): 317–338.

Lafont, Cristina. 2003. Procedural justice? Implications of the Rawls-Habermas debate for discourse ethics. *Philosophy & Social Criticism* 29 (2): 163–181.

Lauer, Helen. 2017. Global justice as process: Applying normative ideals of indigenous African governance. *Philosophical Papers* 46 (1): 163–189.

Lebel, Jean, and Robert McLean. 2018. A better measure of research from the global south. *Nature* 559: 23–26.

Levidow, Les, and Theo Papaioannou. 2018. Which inclusive innovation? Competing normative assumptions around social justice. *Innovation and Development* 8 (2): 209–226.

Lim, Daryl. 2013. Self-replicating technologies and the challenge for the patent and antitrust laws. *Cardozo Arts & Entertainment Law Journal* 32 (1): 131–223.

Manca, Annalisa, Javiera Atenas, Chiara Ciociola, and Fabio Nascimbeni. 2017. Critical pedagogy and open data for educating towards social cohesion. *Italian Journal of Educational Technology* 25 (1): 111–115.

Marion Suiseeya, Kimberly R. 2014. Negotiating the Nagoya Protocol: Indigenous demands for justice. *Global Environmental Politics* 14 (3): 102–124.

Marks, Stephen P. 2011. The neglected human right to benefit from scientific progress: Implications for human development. In *Human Development and Capabilities Association 2011 International Conference*, Den Haag, September 6–8.

McIntyre, Beverly D., Hans R. Herren, Judi Wakhungu, and Robert T. Watson. 2009. *International assessment of agricultural knowledge, science and technology for development (IAASTD): Synthesis report with executive summary: A synthesis of the global and sub-global IAASTD reports*. Washington, DC: Island Press.

McMichael, Anthony. 2017. *Climate change and the health of nations: Famines, fevers, and the fate of populations*. Oxford: Oxford University Press.

Metz, Thaddeus. 2009. African moral theory and public governance: Nepotism, preferential hiring and other partiality. In *African ethics: An anthology for comparative and applied ethics*, ed. F. Munyaradzi, 335–356. Pietermaritzburg: University of KwaZulu-Natal Press.

Millaleo-Hernández, Salvador. 2019. Recursos genéticos y pueblos Indígenas: La tesis de la propiedad cultural indígena frente al dominio público. *Acta bioethica* 25 (1): 51–61.

Muldoon, James. 2019. Academics: It's time to get behind decolonising the curriculum. *The Guardian*, 20 Mar 2019. Accessed 26 Dec 2019. https://www.theguardian.com/education/2019/mar/20/academics-its-time-to-get-behind-decolonising-the-curriculum.

Ottinger, Gwen. 2013. Changing knowledge, local knowledge, and knowledge gaps: STS insights into procedural justice. *Science, Technology, & Human Values* 38 (2): 250–270.

Patel, Raj, and Jason W. Moore. 2017. *A history of the world in seven cheap things: A guide to capitalism, nature, and the future of the planet*. Oakland, CA: University of California Press.

Piketty, Thomas. 2014. *El capital en el siglo XXI*. Mexico, DF: Fondo de Cultura Económica.

Plomer, Aurora. 2012. The right to access the benefits of science and intellectual property rights. In *Biotech innovations and fundamental rights*, ed. Roberto Bin, Sara Lorenzon, and Nicola Lucchi, 45–68. Milan: Springer.

Popa, Eugen Octav, Vincent Blok, and Renate Wesselink. 2020. Discussion structures as tools for public deliberation. *Public Understanding of Science* 29 (1): 76–93. https://doi.org/10.1177/0963662519880675.

Prathapan, K Divakaran, and Priyadarsanan Dharma Rajan. 2011. Biological diversity: A common heritage. *Economic & Political Weekly* 46 (14): 15–17.

Reiss, Julian. 2019. Expertise, agreement, and the nature of social scientific facts or: Against epistocracy. *Social Epistemology* 33 (2): 183–192.

Resnik, David B. 2009. International standards for research integrity: An idea whose time has come? *Accountability in Research* 16 (4): 218–228.

Ricciardi, Vincent, Navin Ramankutty, Zia Mehrabi, Larissa Jarvis, and Brenton Chookolingo. 2018. How much of the world's food do smallholders produce? *Global Food Security* 17: 64–72.

Robaey, Zoë. 2016a. Gone with the wind: Conceiving of moral responsibility in the case of GMO contamination. *Science and Engineering Ethics* 22 (3): 889–906.

Robaey, Zoë. 2016b. Transferring moral responsibility for technological hazards: The case of GMOs in agriculture. *Journal of Agricultural and Environmental Ethics* 29 (5): 767–786.

Rolin, Kristina. 2015. Values in science: The case of scientific collaboration. *Philosophy of Science* 82 (2): 157–177.

Sandin, Per. 2004. The precautionary principle and the concept of precaution. *Environmental Values* 13 (4): 461–475.

Santoro, Daniele, and Manohar Kumar. 2018. *Speaking truth to power-A theory of whistleblowing*. New York: Springer.

Schlosberg, David. 2013. Theorising environmental justice: The expanding sphere of a discourse. *Environmental Politics* 22 (1): 37–55.

Schlosberg, David, and David Carruthers. 2010. Indigenous struggles, environmental justice, and community capabilities. *Global Environmental Politics* 10 (4): 12–35.

Schroeder, Doris, and Carolina Lasén-Díaz. 2006. Sharing the benefits of genetic resources: From biodiversity to human genetics. *Developing World Bioethics* 6 (3): 135–143.

Schroeder, Doris, and Balakrishna Pisupati. 2010. *Ethics, justice and the convention on biological diversity*. Nairobi: United Nations Environmental Program.
Sen, Amartya. 2009. *The idea of justice*. Cambridge, MA: Harvard University Press.
Shiva, Vandana, and Poonam Pandey. 2006. *Biodiversity based organic farming: A new paradigm for food security and food safety*. New Dehli: Navdanya.
Shrader-Frechette, Kristin. 2002. *Environmental justice: Creating equality, reclaiming democracy*. Oxford: Oxford University Press.
Singer, Peter. 1975/2009. *Animal liberation*. New York: HarperCollins.
Solum, Lawrence B. 2004. Procedural justice. *Southern California Law Review* 78: 181–321.
Stilgoe, Jack, Richard Owen, and Phil Macnaghten. 2013. Developing a framework for responsible innovation. *Research Policy* 42 (9): 1568–1580.
Taylor, Jeremy. 2019. Reporting research findings to participants is an ethical imperative. *BMJ* 367: l6324. https://doi.org/10.1136/bmj.l6324.
Thompson, Paul B. 2009. Philosophy of agricultural technology. In *Philosophy of technology and engineering sciences*, ed. Anthonie Meijers, 1257–1273. Amsterdam: Elsevier.
Thompson, Paul B. 2018. Emerging (food) technology as an environmental and philosophical issue in the era of climate change. In *Food, environment, and climate change: justice at the intersections*, ed. Erinn Cunniff Gilson and Sarah Kenehan, 195–211. Lanham: Rowman & Littelfield International.
Timmermann, Cristian, and Zoë Robaey. 2016. Agrobiodiversity under different property regimes. *Journal of Agricultural and Environmental Ethics* 29 (2): 285–303. https://doi.org/10.1007/s10806-016-9602-2.
Tittonell, Pablo. 2014. Ecological intensification of agriculture—Sustainable by nature. *Current Opinion in Environmental Sustainability* 8: 53–61.
Tittonell, Pablo. 2016. Feeding the world with soil science: Embracing sustainability, complexity and uncertainty. *Soil Discuss*. https://doi.org/10.5194/soil-2016-7.
Torrance, Andrew W., and Eric von Hippel. 2015. The right to innovate. *Michigan State Law Review* 2015 (2): 793–829.
UN Committee on Economic Social and Cultural Rights. 1999. *General Comment No. 12. The right to adequate food (article 11) (E/C.12/1999/5)*. Geneva: United Nations Economic and Social Council.
van den Belt, Henk. 2015. Design for values in agricultural biotechnology. In *Handbook of ethics, values, and technological design: Sources, theory, values and application domains*, ed. Jeroen van den Hoven, Pieter E. Vermaas, and Ibo van de Poel, 571–588. Dordrecht: Springer.
van der Ploeg, Jan Douwe. 2010. The peasantries of the twenty-first century: The commoditisation debate revisited. *The Journal of peasant studies* 37 (1): 1–30.
Vermunt, Riël, and Herman Steensma. 2016. Procedural justice. In *Handbook of social justice theory and research*, ed. Clara Sabbagh and Manfred Schmitt, 219–236. New York: Springer.
Villarroel, Raúl. 2013. Ética del desarrollo, democracia deliberativa y ciudadanía ambiental: El desafío global de la sustentabilidad. *Acta bioethica* 19 (2): 189–198.
von Schomberg, René. 1997. Öffentlichkeit als Kontrolle technologischer Innovation. *Analyse & Kritik* 19 (1): 108–125.
Wallack, Michael. 2006. Justice between generations: the limits of procedural justice. In *Handbook of intergenerational justice*, ed. Joerg Chet Tremmel, 86–105. Cheltenham: Edward Elgar.
Walzer, Michael. 1993. Objectivity and social meaning. In *The quality of life*, ed. Martha C. Nussbaum and Amartya Sen, 165–177. Oxford: Clarendon Press.
Wolff, Jonathan, and Avner de-Shalit. 2007. *Disadvantage*. Oxford and New York: Oxford University Press.
Woolston, Chris. 2019. PhDs: The tortuous truth. *Nature* 575: 403–406.
Wray, K. Bradley. 2007. Evaluating scientists: Examining the effects of sexism and nepotism. In *Value-free science: Ideal and illusions?*, ed. Harold Kincaid, John Dupré, and Alison Wylie, 87–106. Oxford: Oxford Univesity Press.

Zakaras, Alex. 2016. Democracy, children, and the environment: A case for commons trusts. *Critical Review of International Social and Political Philosophy* 19 (2): 141–162.
Ziegler, Jean. 2011. *Destruction massive: Géopolitique de la faim.* Paris: Seuil.

Chapter 9
Agricultural Innovation and Restorative Justice: Facilitating Cooperation by Building Conflict Resolution Capacities

Abstract Due to the enormous number of people involved in the continuous improvement of agricultural practices—through tasks as diverse as the selection of seeds with desirable traits for the next harvest to the manipulation of genetic organisms, and from the identification of agroecological principles for reintegrating organic matter to the synthetisation of agrochemicals to fertilize soils—injustices in innovation processes are both foreseeable and unavoidable. Leaving such injustices unaddressed is detrimental for scientific advancement, as scientific enterprises depend on cooperation in order to benefit from the sharing and exchange of samples, observations, feedback and worries. Due to the large number of participants in agricultural innovation and the wide variety of scientific approaches and observation methods, as well as the extensive periods of time some innovation processes involve, it is difficult or even impossible to rely exclusively on conventional legal options to address injustices. Here the idea of restorative justice, which seeks to rebuild social cohesion instead of concentrating on punishment, has great potential. Reconciliation conferences can help to build trust between highly heterogeneous groups—for instance, between indigenous communities and corporate scientists—and suggest alternative compensatory and disciplinary actions.

Keywords Truth-seeking processes · Reconciliation conferences · Historical wrongs · Biopiracy

9.1 Introduction: Why Do We Need Reconciliation?

Political feasibility sets a limit on the success of procedures designed to realize the demands of various forms of social justice. As mentioned in the previous chapter, procedural justice needs to establish a balance between the scope and accuracy it wants to achieve to realize social justice, and the limits in complexity such systems can have to be cost-effective and able to guide people by offering a set of principles that can be realistically acquired (Solum 2004). This major limitation of procedural justice obliges us to deal with major injustices that are left unattended in reconciliatory processes (see Fig. 9.1), so that we can maintain and re-establish good relations

C. Timmermann, *Social Justice and Agricultural Innovation*, The International Library of Environmental, Agricultural and Food Ethics 31,
https://doi.org/10.1007/978-3-030-56193-2_9

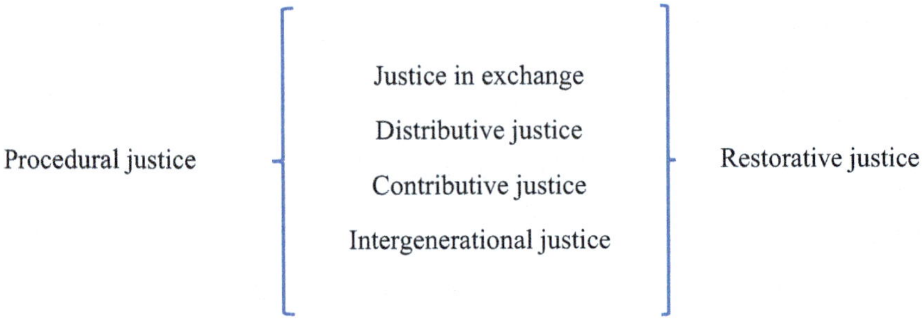

Fig. 9.1 The six dimensions of social justice in agricultural innovation

between people. Social harmony is—with the exception of social cohesion triggered by solidarity in catastrophic events—a prerequisite for exploring and engaging in cooperative enterprises, such as research and development.

For any theory of justice that looks at innovation, it is essential to keep in mind that science is a social enterprise of great magnitude, and as such, wrongdoings due to egoism, malice, a lack of empathy, forgetfulness, an inability to place oneself in the position of others, or ignorance, are very likely to continuously emerge due to the large number of people who participate in such efforts (Jefferson 2006; Koepsell 2016). Nowhere is this as true as it is in agriculture, as working the land is the most common profession, and agricultural systems are the most extensive intervention in the environment. Furthermore, people tend to occasionally break or bend the rules when they are convinced that they are serving a higher goal (i.e. food security). The urgency of improving food systems makes agricultural innovation an area that is highly vulnerable to emotionally charged arguments and fierce disputes over research methods and trajectories. The commercialization of agricultural innovation is also highly lucrative (Korthals 2018), which makes it tempting to abandon principles of good scientific practice. Therefore, worldwide innovation systems urgently need global institutions and locally adapted reconciliation procedures to seek reparations for scientific and corporate misbehaviour—especially biopiracy (Blakeney 2019)—different forms of discrimination (Anderson 2015), and the wilful spreading of false information and misleading advertisements.

It is fundamental for global scientific cooperation that science gains and retains an image of being a just cooperative endeavour, so that people actively seek to share and exchange knowledge, observations, questions, feedback, data and samples, thus contributing to scientific advancement (Langat et al. 2011). In other words, agricultural innovation systems must be deemed worthy of cooperation (cf. Dodds 2005; Ooms 2010). In order to justify the social mechanisms needed to re-establish trust in science and scientific enterprises, I will explore a restorative justice approach to identifying the different elements that require a restorative process in agricultural innovation. A central demand of restorative justice is to correct such injustices. In our case, by actively engaging with wrongdoers and by creating a more receptive research environment for groups and research lines that have historically been discriminated against. This approach requires identifying the type of harm or wrong done, the people who caused (or are causing it), the direct victims and the people who are currently suffering from the long-term effects of such wrongs, in order to build a reconciliatory process (Munzer 2012).

To proceed, I will start by listing some of the gravest problems that are perceived as injustices in agricultural innovation processes, and analyse to what extent these are indeed injustices and why (or why not). I will do so by analysing the problems of biopiracy, the failure to comply with access and benefit-sharing agreements, informed consent procedures, bribery, and causing unnecessary scarcity through socially irresponsible licensing. I will continue by listing the victims of such injustices. I will then conclude by arguing that we should use a framework of restorative justice, instead of relying on retributive justice, in particular by appealing to the need to maintain scientific cooperation and the extreme inequality in power relations between perpetrators and victims.

9.2 Which Type of Injustices Are Occurring in Our Agricultural Innovation Systems?

We must assess the different types of injustice involved that would ultimately require restoration. These injustices may involve different claims and sometimes even obligations. A reconciliation process may not only allow identifying the type of harm suffered, but also the circumstances that led to the harm or allowed it to occur (Braithwaite 2000). The knowledge thereby gained would help in optimizing the preventive processes described in the previous chapter. What kinds of injustice are committed during the innovation processes? We can observe that despite precautionary measures, the demands of all the discussed forms of social justice are regularly not fully met. After discussing violations of commonly established procedures, we will discuss which types of injustice occur, with reference to the forms of justice we have emphasized.

9.2.1 Violations of Standard Procedures

Procedural justice seeks to establish various processes and principles to reach fair outcomes. A violation of these commonly established principles is an injustice to those who were meant to be protected by these principles, and to all other scientific community members who made a substantial effort to abide by these principles in order to improve the reputation of the innovation systems and their institutions in particular. Generally speaking, the most common violation of procedural justice is differential treatment, either by not allowing everyone to benefit from the security of justice,[1] or by undeservingly giving some a stronger voice in democratic decision-making.

Regarding agricultural innovation in particular, a commonly ignored principle is to adequately inform people. Informing people is crucial for procedural justice, as without sufficient information there can be no adequate deliberative process. People should be informed beforehand about the risks and benefits of technologies, so that they can decide on their use. This information needs to be impartial, and given with the intention that it will be understood by those affected. This may require the simplification of technical jargon, illustrations and translation into local languages. The obligation to inform people will depend on the degree to which the technology affects their lives. Any failure to inform the affected parties may create conflicts afterwards, as it would violate principles of self-determination. Efforts to inform people after the event carry the risk that new information becomes available which, if negative, may increase the blame of the non-consulting party even further. It is also difficult for the public to guess in the aftermath what decisions they would have made before the new information became available.

In addition to obligations to inform people about technologies, we have a series of specific obligations. First, as mentioned before, we have the obligation not only to inform, but to also obtain informed consent when using genetic resources or traditional knowledge (Marion Suiseeya 2014). In particular, a failure to inform indigenous communities can erode even further the (often well-founded) mistrust they hold towards governmental authorities and scientists in general. Second, informed consent should also be obtained when land is used as a test site, e.g. as part of an experiment for genetically modified crops or agrochemicals. Products that are still under development should be properly labelled as such.

Lastly, procedural justice extends beyond requiring informed consultations over matters that have deep public implications and concerns, and also demands that democratically taken decisions are indeed implemented. Here the population should be able to retract their consent (or objection) once new information becomes available. There are, however, instances where the protection of minorities and the interests of future generations may force governments to oppose the will of the majority (Bovenkerk 2012), particularly when doing otherwise would imperil securing basic needs (Villarroel 2013). In such cases, governments will have to launch public education campaigns to explain to the citizenship why their will was not respected.

9.2.2 *Market-Based Injustices*

To start, as we saw earlier, social justice problems arise as a consequence of unfair transactions. Isolated and resource-poor farmers are particularly vulnerable to deception and exploitation when they interact with powerful companies and salespersons (Thomas and De Tavernier 2017). In other instances, multinational companies have

[1] I owe this term and the background idea to Martin Luther King's (1963) "I have a dream" speech.

used their power to lobby—often by threatening to relocate to countries that are friendlier to their interests—for regulations that are to their advantage but harm social welfare and environmental protection (Hassoun 2009).

One problem, for example, involves pricing. We are currently witnessing on a massive scale that the sales prices of many Green Revolution-type innovations do not properly incorporate negative externalities, such as soil erosion, carbon emissions, the killing of non-target species and contaminating aquifers (Tittonell 2013; Patel 2013). Short-term benefits in terms of increased yields often bring massive long term costs in terms of lost knowledge, land degradation and pollution (Shiva and Pandey 2006). In some cases, these negative effects are only noticeable after years or decades. Research in conventional agriculture has raised yields considerably, but at the cost of disfavouring research into farming systems that are based on ecological principles. Innovations in conventional agriculture that have major negative externalities must either be prohibited or adequately taxed to finance solutions that compensate these externalities or disincentivize their use. A failure to do so gives innovations that improve yields but have substantial negative externalities a competitive advantage over innovations that are more socially and environmentally beneficial. This advantage is exacerbated by the many subventions that conventional agriculture receives (De Schutter 2009). Compensation has its limits, however: in many cases it is not possible to make technology producers and users pay for all the damage they have caused. For instance, it is particularly difficult to make an inventory of soil degradation and its value (Timmermann and Félix 2019). Soil quality and biodiversity are goods that took millennia to develop and evolve; therefore they are hardly replaceable and any monetary compensation would be arbitrary (El Mujtar et al. 2019)—as the lost good cannot be substituted—and would also bring about issues of intergenerational justice in the long run (as we saw in Chap. 7). At the same time, urban populations have benefited for decades from cheap food prices, despite widespread knowledge of the environmental footprint of agriculture. Benefiting from ecological injustices creates issues of responsibility, which require assuming some of the costs together with technology producers and users. Restorative justice processes could help to find a way in which technology producers and users can work together to revert some of the damages they have caused and develop more environmentally friendly alternatives.

High prices for agricultural innovations not only impede smallholders from benefiting from scientific advancement, but also put them in a position where they will be providing crucial ecosystem services for free that are to the benefit to all. Smallholders contribute to the in situ conservation of landraces, offer refuge areas for key environmentally beneficial organisms and, by not using chemical fertilizers or pesticides (van der Ploeg 2014), allow others to use much more than their environmentally sustainable share without experiencing the full impact that they would if everyone were to do the same. Justice requires that farmers who do not benefit from the advancement of science do not end up providing compensatory ecosystem services for others without receiving any reward for their effort.

Another problem involves the acquisition of smallholder knowledge. According to international regulations, just giving credit and money for innovations is not enough.

When genetic resources are involved, rewards need to be fair, agreed on mutual terms and prior informed consent needs to be obtained (Brody 2010). Following the Convention on Biological Diversity of 1992, the most accepted practice for acquiring biological resources and traditional knowledge is to come up with access and benefit-sharing agreements (De Jonge 2011). Both parties need to agree on the terms, the conditions need to be fair, and both parties should abide by the stipulated conditions after the agreement was reached (Schroeder and Pisupati 2010). In any agreement and exchange procedure, it is crucial for informed consent to be obtained beforehand, so that both parties have adequate knowledge of the conditions and effects of their exchange (see Chap. 4). Issues of intergenerational justice need to be considered here as well. If a company offers a generation within a community luxury goods—such as televisions or the latest smartphones—for the full exploitation of their ancestors' traditional knowledge, then we may question the fairness of such deals, even when they meet all requirements of access and benefit-sharing principles. Despite inevitable accusations of paternalism, offering common or public goods in exchange should be the preferred strategy, as this will also allow subsequent generations of the same community to benefit from these transactions.

A very different problem is the artificial scarcity that intellectual property regimes bring about. Unnecessary scarcity may cause significant harm and deteriorate social relations. As mentioned before, intellectual property allows rights holders to engage in licensing practices that will create an artificial scarcity of a product that would otherwise exist in abundance (Sterckx 2005; Timmermann 2017). By pricing an innovation highly, access is limited to those with abundant financial resources. Agricultural inputs to combat pests or improve soil fertility, or innovations that allow more efficient irrigation, often need to be purchased. In times of drought or pests, richer farmers can gain a considerable advantage over poorer farmers, especially when these innovations are expensive, often with the consequence that poorer farmers end up selling their land to richer farmers at low prices. Such scenarios will become more frequent as the effects of climate change worsen, and will make food production more difficult. Considering historic emissions, innovations for adapting agriculture to climate change raise major justice concerns (Biddle 2016, Timmermann and van den Belt 2012). Restorative justice processes will have to deal with unfair or disproportionally burdensome absorptions of the costs of innovation. This is particularly important when it comes to the distribution of costs, in order to absorb the global burden of climate change adaptation and mitigation and end hunger.

Lastly, unfair market competition also has a major social cost. In theory, competitive markets favour the survival of efficient inventors. Yet monetary resources can also be used to block cost-effective inventions, especially through the acquisition of patents, in order to sell more lucrative products. This has been described as using patents in a "destructive form" (Schneider 2010; Heins 2009) instead of a "creative form", as it hinders follow-up innovation and commercialization, by imposing licensing conditions that are commercially unviable or socially unacceptable. This again leads to path dependency once these more lucrative inventions become industry standards, putting research lines that were originally more socially useful in a position of disadvantage. Many such innovation paths need to be reassessed in relation to their role in contributing to social welfare.

9.2.3 *Unfairness in the Allocation of Research Funds*

Agricultural innovation is one branch of science that has suffered extensively from scientific impartiality. In particular, innovations that can be reproduced on the farm have been widely ignored by the scientific community. This is particularly worrying, since the global poor are unlikely to benefit from the advancement of science unless these innovations can be adopted and replicated using locally available spare resources, thus raising issues of social justice (Gupta 2010).

One stream of agricultural innovation that has been particularly neglected is agroecology: a farming system that aims to work with nature to build sustainable food systems and thereby minimize the need for externally produced inputs (Tittonell 2013; Vanloqueren and Baret 2009). There are several accusations to the effect that conventional agriculture owes its perceived superiority to enjoying privileges that have no full scientific backing, or merely to strategic business and political interests. In other words, a research trajectory that initially creates a commercial advantage is eventually seen as best practice from a scientific perspective. For example, hybrid maize varieties have had decades of research advantage over open pollinating maize varieties, due to the central factor that the former give much more control to seed companies (Kloppenburg 2005). Academic incentives make it particularly attractive to cooperate with industry, concentrating research attention in areas where innovations can be commercialized through the use of intellectual property rights (Stephan 2012). Agroecology is also negatively affected by academic career assessment tools, due to the length of the field studies that are needed to attain publishable results (Geertsema et al. 2016).

In addition, an overemphasis on the harvest yields of specific crops has led other dimensions to be ignored (Thompson 2017). There are a number of negative and positive externalities that are given insufficient consideration in agricultural politics when drafting scientific agendas. It is now urgent to re-evaluate the positive and negative long-term effects of agricultural innovations. While, for example, improving methods for capturing carbon in agricultural land have substantial benefits for food security and the halting of climate change (Altieri et al. 2015), the use of several agrochemicals had to be prohibited after harmful effects on both human and animal life were proved (Kroma and Flora 2003). Conventional agriculture also needs a massive amount of cheap fossil fuels for the production of fertilizers (Tittonell 2013). It is also dependent upon a large number of migrant workers, due to the uniformity of harvesting periods, raising major social justice issues (Loo 2014). This has led to accusations that conventional agriculture has benefited from an unfair advantage, and that this advantage is maintained by continuously ignoring the major negative externalities that are in dire need of being redressed. This failure to integrate ecologically superior food production methods into common knowledge systems leads to the injustice of being unable to live in harmony with nature.

As the diversion of public research funds towards commercial needs was caused by the aggregate actions and omissions of a large number of scientists, company managers, farmers and politicians, a restorative justice process is needed. Such a process needs to identify the major research lines that were unduly discarded: those whose knowledge was left aside and those who took unfair advantage of this outcome. In this way the actors involved can work in common towards a solution.

9.2.4 Unfair Competition of Ideas

As mentioned in Chap. 3, Robert Merton is probably the most influential science theorist who has advocated adhering to the norm of scientific disinterestedness, that is, to follow only one's scientific curiosity in the quest for truth (Merton 1942). Pride in one's personal contribution to science, strong feelings of belonging to a specific school of thought, commercial interests and state agendas are all seen as obstacles to and interferences in the advancement of science.

In addition to the just mentioned commercial interests in unfairly favouring one line of research over another, in the history of science and innovation there have been a number of cases where this partiality was due to personal and state interests, despite scientific evidence or recommendations favouring other paths. While some such interests are indeed very laudable—as are safety concerns to protect humans and the environment, or threats to world peace—others are not, as is the case when technologies and methods are only favoured to maintain a nation's or even a university's dominance.

There have been far more aggressive ways to impose certain lines of research. As mentioned in the previous chapter, a factor that has had a strong negative influence on agricultural innovation is corruption: either by directly bribing public officials or research institutes, or by using positions of power to favour less qualified or biased scientists, or research lines that accord with personal gains or convictions. In addition, there are cases of data fabrication and falsification. Cases of scientific malpractice and corporate corruption of research are particularly difficult to identify in the aftermath, as it is extremely costly to systematically revise accepted scientific evidence. The scientific community, alongside scientific publishers, needs to maintain a hard line in confronting researchers who deviate from good scientific practice. A restorative justice approach demands that we also recognize the affected scientists who were harmed by such practices as victims, taking into special consideration that they are likely to suffer from multiple discriminations in many areas of life, and demand that corrupt scientists formally apologize to the affected researchers.

A very large proportion of universities still retain the liberty to discriminate against scientists on the basis of faith, even when they receive governmental funding for research: ultimately limiting the pool of competitors for positions. The recurring failure to choose the best qualified scientists or the most sustainable research programmes has catastrophic results for scientific advancement and public welfare (cf. Ceva and Ferretti 2018). The failure to ensure equality of opportunity discourages whole population groups from engaging in science, obliges under-represented groups to leave the country, and facilitates nepotism, which disfavours the introduction of new ideas and changes of scientific paradigm.

In addition, there is discrimination without any ethically acceptable justification. The fact that unjustifiable discrimination is a form of harm is self-evident. It affects the scientific community in many forms, including discrimination on the basis of gender (most massively against women), age, religion or lack thereof, race, and ethnic background, in particular identifying as a member of an indigenous community. Yet the case of justifiable discrimination is somewhat tricky to assess. Favouring a certain intellectual tradition, school of thought, method or institution over others saves resources. Even though such prioritization is not a very good filter, some sources do usually provide a minimum standard of quality. For example, many publishing houses and organizations have a long record of carefully screening their publications, which reduces the likelihood of finding information of poor quality in these venues. A major problem of using well-known sources as a form of filter, however, is that we fail to acknowledge the historical discrimination that many knowledge providers have suffered and thereby continue to give established sources an unfair head start. Relying on the same sources also contributes to epistemic poverty, thereby slowing down scientific advancement.

The problem of rectifying injustices that result from discrimination, is that beyond clear-cut cases of explicit discrimination, it is sometimes difficult to identify who is responsible for discriminatory practices and at what stage a practice can be objectively counted as discriminatory. Let us examine some of these difficulties. To start, we have inherited certain practices and infrastructure that make it difficult, or even impede, people from participating in scientific projects. For example, old and often cheaper infrastructure may impede access for people who depend on wheelchairs for their mobility. Some courses can be carried out late at night in buildings that are cheap to rent, but located in areas that may disproportionally dissuade women from participating due to security concerns, even when those courses become financially accessible for a wider population segment. While constraints on resources may oblige us to make difficult choices about who will end up being excluded, we need to pay special attention to ensuring that the same people are not always negatively affected. It is clear that some groups—particularly rich men—are rarely excluded at all. What are the responsibilities of those people who continuously benefit from the current social arrangements? Are there any obligations to fight, as individuals, such exclusive practices? Adhering to the principle of equality of opportunity obliges us, as individuals, to assist people in less fortunate circumstances.

The difficulty of identifying some discriminators and cases of discrimination makes deterrence by punitive measures insufficient by itself. Innovators need to be taught about the injustice of discrimination and the benefits of inclusive research environments, so that they will have an interest, as individuals, in defending and promoting cognitive diversity. Somewhat different is the situation with clear-cut cases that can be sanctioned by social stigma and/or institutional and legal sanctions. An emphasis on restorative justice does not imply that we do not need legal sanctions for severe cases (Poama 2015), such as hate speech, sexual abuse, and the diversion of public funds for private purposes.[2]

Another form of harm we can list is the failure to recognize members of certain groups as peers (Fraser 1998). Here again, this type of differential treatment can be observed in multiple forms. For example, treating an invention as coming from a certain collective—e.g. indigenous communities or smallholders—in a world where "reputable" authors are recognized as individuals, fails to recognize knowledge providers from certain communities as individuals who act autonomously, have distinct forms of inventiveness and creativity, and overcome hurdles independently. This is particularly problematic when traditional communities are perceived as somewhat static collectives, as mere stewards of knowledge (Robinson 2008), while individual innovators are seen as dynamic, and as producing innovations tailored to current needs.

The failure to recognize certain people as the authors of inventions often goes hand in hand with depriving those people of economic remuneration (Deplazes-Zemp 2019). This leads to the type of situation that recognition theorists have often labelled as "adding injury to insult". Recognizing authorship often leads to economic remuneration, and paying inventors often leads to recognition for their work. There is an economic incentive to not recognize people as authors, as sharing credit entails sharing royalties. Such practices often also have the effect that they discourage such communities from becoming entrepreneurial and seeking compensation for their innovations (Gupta 2006).

The extent to which we should condemn the failure to recognize someone depends on the extent to which we understand there to be a special obligation to recognize someone or someone's effort. We usually think of people as deserving when the benefit they provided was foreseeable, sizeable, time and resource-consuming, took perseverance and skill, or involved risks or unpleasant work (Annis and Bohanon 1992), as opposed to when they attained a beneficial outcome by sheer coincidence. Yet, arguably, this alone does not entail an obligation to individually "provide" such recognition. People could demand recognition if we usually granted recognition to others for the very same reasons, or if we personally demanded recognition for our own contributions. A repeated failure to recognize the effort made by specific groups while recognizing the same type of effort when made by others would amount to a form of discrimination. A clear academic case occurs when white male doctoral candidates are repeatedly praised for their intellect, while their female colleagues are congratulated for their beauty, and their male colleagues of colour are congratulated for their athletic prowess. A similar situation arises if we somehow draw an unjustifiable distinction between those who we recognize as our peers and those who we do not. In other cases, we could say that decency requires to recognize efforts

[2]It has been documented that restorative justice conferences even offer very good results for major crimes, especially when it comes to identifying perpetrators. See Braithwaite (2000).

which we depend upon or benefit from, irrespectively of whether we have requested these services or not. Due to historical discrimination and inequality of opportunity, we may fail to acknowledge privileges and hurdles if we provide everyone with the same recognition (Bascara 2016). This would fail to recognize the difficulties that some have had to endure and the advantages enjoyed by others (Schweiger 2014). We need to recognize the effects of different life experiences and the individuality of hardship. In practice, one would have to live as a hermit, without any expectations from anyone, in order to safely claim that one does not owe anyone recognition.

When it comes to innovation itself, blindly privileging a certain school of thought may lead to the so-called "path dependency" in research, which makes it extremely costly to change back to a more sustainable research trajectory, ultimately forcing society to continue to do research on a less beneficial path (Vanloqueren and Baret 2009). We must formally recognize when a research line has received an unfair head start over another (Coolsaet 2016).

While, as a general rule, one should hire the people and follow the research lines that are in the general public's best interest, impartiality is not always the best strategy (Metz 2009). There are background situations where it might be in the general public's best interest to support, for instance, affirmative action policies that give people or research lines that have been unjustly historically disadvantaged additional assistance to ensure a more representative epistemic diversity. The success of such policies will not only depend on how they are implemented, but also on how well the need for and justification of such policies is understood by the general public, especially by the parties directly affected. Educational campaigns will be needed to teach people about the difference between affirmative action and differential treatment that meets only private interests. Failing to do so will likely result in an inability to distinguish between someone being given an additional advantage in view of historical circumstances from someone being given the same advantage merely for being part of a community.

9.2.5 Long-Term Social and Environmental Sustainability

During the twentieth century, the most urgent publicly defended target that was set in global agricultural innovation agendas was to reduce one particular harm: hunger. Despite the fact that this major social goal is still a top priority and necessity (Lappé et al. 2013; Hickel 2016), it is now widely acknowledged that achieving food security should not and does not have to be pursued at all costs (Oteros-Rozas et al. 2019; Ickowitz et al. 2019). Agricultural innovation streams continue to deviate considerably from a number of long-term sustainability targets, ranging from nature conservation to losses in food sovereignty.

One of the most pressing issues of concern is the conservation of nature. Conventional agriculture is leaving an enormous environmental footprint, by releasing toxins into the environment, killing non-target species, and emitting greenhouse gasses on a massive scale (McIntyre et al. 2009). The enormous negative impact that conventional agriculture is having on the environment is condemned by appealing to the instrumental value of nature, as a provider of valuable ecosystems, and also to its intrinsic value, as something that people today and in the future appreciate for its own sake. Viewing the environment solely as a distributive good underestimates and misconceives our ties to and views of nature (Jamieson 1994). Over-incentivizing conventional farming systems may impede individuals and communities from living in mutualistic relationships with nature. As mentioned earlier, being able to live in such relationships is of essential importance for some cultural and philosophical values (Rozzi 2012). Our current rate of environmental destruction is leading to an ecocide: killing ourselves and others (Neira et al. 2019). This would require restorative measures, to once again allow certain forms of livelihood that are endangered by our current massified and near-sighted consumerist behaviour. Currently, there is an increasing awareness that food systems do not need to be environmentally damaging, as ecological intensification is already an attractive way to produce food in symbiotic relationships with adjacent ecosystems and also to provide a living space for much more wildlife and domesticated species within farms (Tittonell et al. 2016). Instead of fighting nature, food production systems can be redesigned to work with nature to produce food (Stojanovic 2019; Blok and Gremmen 2016), which allows us to provide room for much more flora and fauna, in both rural areas and cities.

It is somewhat tricky to assess the case when technology leads to dependency over the years or decades, leading to a loss in food sovereignty. Many Green Revolution -type technologies show their potential only when used in combination with other technologies; for example, some seeds will need better irrigation systems, fertilizers and pesticides. Higher yields can usually only be maintained when the seeds are used in combination with these other inputs. In cases where innovators were open to this dependency, and people knowingly decided to accept these disadvantages on the basis of the added benefits, we would not condemn as much the sales of such technologies. It is different when innovators market their technologies without being clear about the resulting situation of dependency and the difficulties involved in returning to past food production methods. A short-sighted adoption of technologies may result in a massive loss of tacit knowledge of farming methods that allow self-sufficiency. This loss of one type of knowledge in favour of another may produce a loss of independence: something that future generations may condemn. We may need to develop policies for re-empowering people to produce their own food without depending on externally produced inputs (Carlisle et al. 2019).

Another form of injustice occurs when the negative externalities of an invention hinder others from working with a particular farming system, as is the case with the effect of pesticides and agrochemical spills on agroecological farms. Court cases and regulations, such as the decisions to favour users of genetically modified seeds in genetic contamination disputes (Robaey 2016), reveal one farming method and research line to be privileged over others.

9.2.6 *Multiple Injustices*

Perhaps the most prominent forms of injustice in agricultural innovation are those that fall into the category of unconsented bioprospecting—or as it is more commonly referred to in advocacy, "biopiracy"—that is, failing to properly acknowledge the intellectual contribution of people towards innovations in biological materials, or not giving the stewards and producers of such innovations a fair financial reward (Hamilton 2008; Blakeney 2019). Usually the people who are cheated out of their financial reward and intellectual recognition come from indigenous communities, or are smallholders or (also) scientists with few resources, or come from historically under-represented or discriminated groups. Positions of advantage are often used to undermine the rights of less privileged inventors, by abusing bargaining positions, and sometimes even by making threats. In other cases, the enormous bureaucracy of host countries, including instances where multiple ministries of the same country claim to have negotiation rights over genetic resources, makes the legal acquisition of usage rights enormously time-consuming or prohibitively expensive, with the consequence that companies opt for illicit acquisition or abandon the research altogether (Conde Gutiérrez 2018).

Biopiracy cases are often not discovered at all, and sometimes only after a considerable amount of time. Yet, due to the multiple forms of justice affected by biopiracy—from economic interests to recognition as peers (Deplazes-Zemp 2019)—it becomes particularly important to engage in reconciliatory measures. Traditional ecological knowledge is of great help to formal scientists, as it narrows down immensely the search for biological substances with potential use (Blakeney 2019, Herrera Vásquez and Rodríguez Yunta 2004). After indigenous communities have identified uses of particular biological substances, the work of scientists is concentrated on testing those components for efficacy and exploring industrial applications, saving massive time and resources on wide-scale bioprospecting.

9.3 Who Is Wronged by Innovation?

Injustices in innovation process are not suffered by all equally, and failing to acknowledge group-specific vulnerabilities is an injustice in itself. To address injustices that need restoration, we first need to identify who has been wronged or suffered a systematic disadvantage. Here we can distinguish four major groups: (1) innovators who are harmed as individuals, (2) innovators who are harmed or neglected for being (or not being) members of a specific community or group, (3) people who have special interests in certain types of innovation, (4) people who have an interest in justice, and (5) innovators whose knowledge has been made useless (instead of superseded). Let us differentiate these five partially overlapping groups.

The first group consists of innovators who have been wronged independently of whichever group they are identified with. Here we find research partners who have been deprived of their fair share or not been properly acknowledged. We can also list farmers who were not properly informed when they participated in an experiment or were part of a study. The distinctive feature of this group is that it consists of people who suffer these injustices not primarily because of who they are, but instead simply due to the neglect, opportunism or ill will of others. These wrongs are directed at them as individuals, and not necessarily as members of any specific groups.

The second group includes innovators who were most likely to be wronged because they are or were members of particular groups of people. The extent to which members of this group are prone to be wronged varies considerably. For example, indigenous communities in countries that have a weak record of protecting their interests are particularly vulnerable to biopiracy. Younger scholars with weak ties to their institutions are much more vulnerable to being plagiarised than more established scholars of the same institutions. Younger female scholars are additionally much more vulnerable to sexual harassment than their male colleagues and senior faculty members. In addition, racial discrimination is still a major hurdle for more inclusive educational and research environments (Gomberg 2007). Of course, people can be subject to multiple forms of discrimination at the same time, as scholars of intersectionality have documented in great detail (Byrd et al. 2019). Often people from such groups are unlikely to receive any justice when they denounce the wrongs they have suffered. In many cases, the victims do not even bother to complain, as they have little or no trust in the system, or fear that the probability of severe retaliation is much higher than the probability of finding any kind of justice worthy of that name.

In the third group, we find those people who have an interest in particular research lines receiving due recognition. This group includes a wide range of people. Many research lines have been discredited for not leading to commodities, or even for mostly being developed by people who have been or are still being discriminated against. Here people are wronged through the loss of traditional ecological methods, seed varieties and local culture that have been abandoned due to false beliefs or misplaced incentives. People do not need to have a special attachment to the knowledge or research line in question—for instance, through their centre of life, community, religion or culture—to be wronged. Having an interest in the discredited knowledge or research line suffices. Of course, personal or communal attachment aggravates the experience of injustice.

In the fourth group, we find those people who have a strong interest in justice and have been wronged by having to work in unjust environments (even if they themselves do not suffer discrimination). We find in this group both average citizens and the well-intended scientist who wants her community to be seen as just. Exploitation and discrimination taint the image of the scientific community, and therefore also the images of its individual members. While we cannot compare this kind of harm to the kind that is suffered by people who are directly affected by injustices, we should not underestimate how injustices within an institution can erode people's sense of belonging, as well as social cohesion among group members and incentives to give one's best performance (van Dijke and De Cremer 2016). When individuals destroy, through their wrongful conduct, the positive image of an institution that others have worked hard to build, the former also harm the latter, and not only those who are directly affected. This is particularly problematic in cases where institutions rely on volunteer work or when people are poorly compensated, as in citizen science projects. Agricultural innovation relies massively on unpaid work, as demonstrated by peer reviewers, agricultural extensionists, farmers who report side effects or user innovations, and formal and informal teachers and supervisors, among others. It would be a major drawback for scientific advancement if people started to demand remuneration for every type of contribution to science and innovation projects.

The fifth group mostly includes holders of traditional ecological knowledge. When the natural habitat of traditional knowledge holders, such as rainforests, is destroyed or when a species is killed nearly to extinction, as happened to the North American buffalo, the knowledge that people possess about those habitats and species becomes practically useless. Since traditional knowledge is mostly passed on orally and often involves tacit knowledge, such knowledge tends to be lost when its possessors cannot pass it on from person to person in traditional conditions. Grandparents cannot teach their grandchildren about the uses of certain species if they cannot find them, or they no longer grow in their natural habitats. In other words, the possibility of passing on indigenous cultural knowledge is dependent on the continuous presence of the relevant biological material in nature (Rozzi 2012). Climate change is another major threat to traditional ecological knowledge. Changes in weather conditions can result in continuous crop failure and in the migration and extinction of species about which indigenous communities have special knowledge, making past knowledge useless due to climatic changes. In more extreme cases, it can oblige people to become climate refugees, and lose all interaction with their traditional lands and waters (McMichael 2017). It may also force people to stop using heritage seed varieties when radical changes in climate make those varieties unsuitable. Since outsiders are the main historic and present contributors to climate change, we can indeed talk about a threat that is inflicted upon those people and thus deserves some type of remediation (Figueroa 2011; Iliescu et al. 2018). This case is different from the one where a technology is superseded by more efficient methods, where past knowledge loses only its relative value and not its absolute value.

These distinctions in no way imply that these different people suffer comparable harms or that they are similarly affected. A doctoral researcher whose results from years of work are plagiarized will suffer much more than a senior professor facing the same fate. Some cases of biopiracy increase demand for a variety to the extent that it becomes financially inaccessible for the communities who stewarded it for centuries, resulting in a double harm: loss of opportunities to gain income and increases in prices (or scarcity) (Desai 2007). Sometimes discrimination and lack of opportunities are massively discouraging, with the result that certain groups abandon science or never even consider it as a career. While these harms and wrongs are of various degrees, they all create disincentives to collaborate with scientific institutions. The aggregation of these disincentives and worries reduces willingness to cooperate (cf. Zhang 2017) with scientific institutions and erodes public trust in science. We therefore need to identify these wrongs so that they can be properly addressed.

9.4 Why Restorative Justice?

Restorative justice builds upon a commitment to restore past relations. This does not mean that the exact social relations of the past need to be restored. In many cases this is simply impossible, as the case of colonization shows. What we can do as a society is to restore social cohesion by confronting the past, recognize errors and wrongdoings, and explore and encourage new types of social relation and interaction (Cohen 2016). People do not need to relate in the same way they used to do, but it is important that they are willing to seek and explore ways of cooperating with each other. In this spirit, restorative justice seeks to establish harmony among victims and wrongdoers. Instead of seeking retribution—in the sense of demanding an eye for an eye, or some type of punitive measure grave enough to deter wrongdoings, or sometimes pursued just for the sake of punishment—restorative justice measures are designed to deter people from committing wrongs and to make it possible for wrongdoers and victims to once again share a space together, or even to cooperate (Braithwaite 1999; McGeer and Pettit 2015). Restorative justice seeks conviviality. In contrast, the concept of retribution goes back to the idea of paying back one's debt: in some traditions it demands that wrongdoers pay their debt to the social institution of justice. Here, even if no specific person is there to receive justice, punishment is sought in the name of the abstract entity of justice.

There are different ways to organize restorative justice conferences, depending on the offence, the people involved and the legal environment. The formality of these conferences varies as well. Depending on the legal framework, they can replace courts or only be allowed for minor non-criminal offences (Poama 2015), for example in the workplace and educational institutions.

We can think of conferences that aim to re-establish trust after cases of biopiracy, sexual misconduct in research institutes, bullying, discrimination and authorship disputes. There are a number of reasons why restorative justice measures are particularly attractive when addressing injustices in agricultural innovation systems.

First, it is crucial for individuals to cooperate with innovation systems not because they have to, but because they want to. It would be impossible (and even undesirable) to monitor in any detail who is hiding away useful information and samples, and thus slowing down scientific advancement and the improvement of social welfare. Since restorative justice measures seek a dialogue with a much larger group, its effect extends to much wider circles. Its aim is not to punish individual perpetrators in order to deter people from committing wrongs, but rather to re-establish trust and willingness to explore cooperation opportunities.

Second, intellectual property rights are usually defended through courts and, especially within this system, stronger parties have a substantial advantage. They can hire the best legal specialists, and can sometimes even prolong court cases in the hope that their counterparts will be unable to continue paying their legal expenses. For such reasons, when highly unequal parties are in dispute, restorative justice measures tend to offer better results on average (Braithwaite 1999).

Third, the victims of injustices regarding innovation are often a dispersed group, with the effect that it is unclear who was harmed and to what extent. A problem for theories of punishment, including retributive justice, is the need to identify a specific wrongdoer. In a social enterprise, as in agricultural innovation, where ideas and genetic resources are taken from a variety of sources and considerable work is added in each step, it is difficult to determine who the exact wrongdoers are. This makes it impossible in practice to seek justice for all who are affected. Larger reconciliatory processes and scholarship dedicated to examining historical injustices can, however, identify relations that facilitate a disproportional distribution of advantages and vulnerable positions, so that such relations can be recognized and redressed (Grey 2019).

Fourth, some of the injustices committed in innovation processes are hard to identify individually. Many actions and behaviours have a negative effect only in aggregate, or in contrast to past or present experiences (Anderson 1995). Individual jokes may add up to be perceived as bullying, and smaller acts, like prizing a colleague's good dress sense, are unlikely to be well-received if combined with repeatedly ignoring her or his scientific merits. Reconciliation processes are especially apt for teaching people to put themselves in the position of others, to learn about other cultures and perspectives, to identify specific needs, aversions and fears, and gain sensitivity thereof (Drewery 2016).

Fifth, there is widespread disagreement about how far responsibility applies in cases where people neglected or omitted to take certain precautions, in contrast with cases where it is clear that the harm was caused wilfully (Miller 2017; Bratu 2017). Since restorative justice has a strong outcome orientation, it aims to reach a consensus on the identification of harm, its causes and victims, with the goal of (re)building trust among all involved parties. Since this consensus is reached collectively rather than unilaterally, it has a higher chance of being accepted.

Sixth, corrective justice measures in particular call for perpetrators to compensate their victims. This requires not only that both are clearly identified, but that both are alive. Many injustices resulting from agricultural innovation occur through lengthy periods of times, often continuing to be unaddressed after one or both parties are deceased. Restorative justice can better address issues between loosely bound group members.

There are some cases that are particularly challenging for restorative justice. For a case to be condemned by a conference among the stakeholders, the majority of the people involved have to agree that the act under discussion is unambiguously wrong (Braithwaite 1999). This will not be possible with many questions concerning innovation and ownership, as there is no consensus on when one can legitimately exclude others (e.g. innovation to secure basic needs), or on what counts as legitimate rent-seeking (leaving people empty-handed despite need). It would be extremely difficult to reach a consensus on what amounts to legitimately profiting from vital technologies. Yet the fact of such matters becoming part of a public debate already brings people together to discuss different perspectives and impressions, to become knowledgeable about the subject matter and reconsider their positions (Bovenkerk 2012), which is already part of a reconciliatory process.

It is also problematic that, from a social perspective, the wrongdoing is sometimes partially compensated by an improvement of the illicitly acquired object. While we may condemn the methods used to develop an innovation, such as exposing people without their consent to agrochemicals to test their toxicity, we are often not as strict, or at least less consistent, in condemning benefitting from the results obtained from such research. Here we can also find less severe cases, such as the plagiarizing of forgotten texts, or of very inaccessible editions, and developing through them some key ideas. While these acts are still malpractice, it is somewhat difficult to say that someone has been harmed by this act and even more difficult to say that society does not benefit from it. Despite the fact that unethical research is generally condemned by society, there is no consequent effort to delete or ignore findings reached through illicit or even inhuman methods. This leaves open a significant incentive for scientists who accept the likely outcome that they will be remembered as monsters, corrupt people or simply erased from textbooks, but carry on under the firm belief that their actions will contribute to the greater social good by saving many more lives in the future. In other words, they act with the belief that the end justifies the means, even when they know they will be personally penalized. Here we may ask ourselves how strict we, as society, should be in maintaining deterrence. Even though we would need exact empirical evidence, intuition suggests that we would be better off with a system that condemns all types of scientific malpractice.

Finally, we must note that many wrongs committed in scientific endeavours are committed while doing something that most would consider a supererogatory act (an exception to this position was discussed in Chap. 6). Harsh penalties may act as a disincentive and are morally complicated, as we occasionally benefit from such malpractices.

9.5 Remediation Efforts

Assuming that these injustices and unfair advantages indeed call for remediation, how could we restore these past and current injustices? Much of the literature on redressing historic injustices focuses on situations where groups of people are the beneficiaries of restoration measures.

Ecological intensification based on traditional knowledge has in particular been subject to historical discrimination, as the knowledge provider is not acknowledged as a provider of socially (and environmentally) useful information (Desai 2007). This could be interpreted as a form of testimonial injustice, as a particular group of innovators is given less credibility (Fricker 2007). Similarly, the way much of this traditional knowledge is being documented and passed on may make it subject to a form of hermeneutic injustice (ibid.) from the scientific community, as it regularly fails to acknowledge the scientific merit of knowledge depicted in non-standard scientific terms, language and outlets. In our case, this framework has great potential to address the injustices that most indigenous communities and smallholders have suffered regarding their intellectual contributions to sustainable agriculture (Waldmueller 2015). Yet it has a major limitation: by focusing on the discriminated person it does not adequately reflect the injustice of not adequately incorporating, as a society, a range of ideas into our shared knowledge systems.

Now, several problems arise when trying to apply restoration measures meant for people to ideas. First, while we could question the distribution of social goods when they are not distributed more or less equally among different group members, e.g. when men have much more than 50% of faculty positions, it is much more difficult or simply impossible to calculate how many ideas from certain sources, or of certain types, should be supported, e.g. the amount of funding that should go to the assessment of traditional knowledge. We lack a generally agreed point of reference. Even if we had such a reference to draw comparisons, scientific approaches and major social welfare concerns may lead us to regularly abandon under-represented research lines if they do not show sufficient potential social or ecological value.

Second, justice requires every human being to be adequately represented and treated with dignity. The case is different with ideas. While it would be an injustice to completely ignore or "delete from the textbooks" every trace of a civilization, the limits of human intellect make it necessary to discard ideas – constantly. Some ideas, especially inventions, are superseded, while others become redundant or have merely historical value. Moreover, social interests call for placing a higher value on ideas that may enhance social welfare, and there are strong arguments for at least making access to some types of ideas more restrictive, e.g. inventions for weapons. There are good reasons to treat ideas instrumentally, to improve social well-being and advance science.

9.6 Afterthought: Restoration for Whom? for What?

Lastly, we may wonder to whom or to what we owe restorative justice. Do we owe it to the memory of the ancestors who developed these techniques? To the present and future generations who may benefit from such knowledge? Or do we owe it to a social order that abides by principles of justice to redress such wrongdoings? To answer this question, we need to be able to agree on a question as fundamental as why we have punishment. This is something I cannot answer, as it is generally a question on which there has been little agreement throughout history (Hassemer 2009).

Maybe, depending on how environmental protection efforts evolve over the coming decades, the largest harm done by this discrimination has been to deprive us of the possibility of living in harmony with nature. We are losing the ability to live in mutualistic relations with nature, and it is becoming more and more difficult to avoid living in parasitic relationships with nature. Here we are not only impeding a particular lifestyle, but also, especially for many indigenous communities, this environmental degradation is impeding people from living according to their cultural values and worldviews. It is also endangering human survival itself (Neira et al. 2019), when we look at the carbon emission from agriculture and the massively underused opportunities of using food production facilities as carbon sinks.

References

Altieri, Miguel A, Clara I Nicholls, Alejandro Henao, and Marcos A Lana. 2015. Agroecology and the design of climate change-resilient farming systems. *Agronomy for Sustainable Development* 35: 869–890.

Anderson, Elizabeth. 2015. Feminist epistemology and philosophy of science. In *The Stanford Encyclopedia of Philosophy* ed. Edward N. Zalta. Standford: Standford University.

Anderson, Elizabeth S. 1995. The democratic university: The role of justice in the production of knowledge. *Social Philosophy and Policy* 12 (2): 186–219.

Annis, David B., and Cecil E. Bohanon. 1992. Desert and property rights. *The Journal of Value Inquiry* 26 (4): 537–546.

Bascara, Rachelle. 2016. Compatriot partiality and cosmopolitan justice: Can we justify compatriot partiality within the cosmopolitan framework? *Etikk i praksis-Nordic Journal of Applied Ethics* 10 (2): 27–39. https://doi.org/10.5324/eip.v10i2.1921.

Biddle, Justin B. 2016. Intellectual property rights and global climate change: Toward resolving an apparent dilemma. *Ethics, Policy & Environment* 19 (3): 301–319.

Blakeney, Michael. 2019. Remedying the misappropriation of genetic resources. In *Intellectual property issues in microbiology*, ed. Harikesh Bahadur Singh, Chetan Keswani, and Surya Pratap Singh, 147–161. Singapore: Springer.

Blok, Vincent, and Bart Gremmen. 2016. Ecological innovation: Biomimicry as a new way of thinking and acting ecologically. *Journal of Agricultural and Environmental Ethics* 29: 203–217.

Bovenkerk, Bernice. 2012. *The biotechnology debate: Democracy in the face of intractable disagreement*. Dordrecht: Springer.

Braithwaite, John. 1999. Restorative justice: Assessing optimistic and pessimistic accounts. *Crime and justice* 25: 1–127.

Braithwaite, John. 2000. Restorative justice and social justice. *Saskatchewan Law Review* 63: 185–194.

Bratu, Christine. 2017. Korporative und kooperative Verantwortung. In *Handbuch Verantwortung*, ed. Ludger Heidbrink, Claus Langbehn, and Janina Loh, 477–499. Wiesbaden: Springer.

Brody, Baruch. 2010. Intellectual property, state sovereignty, and biotechnology. *Kennedy Institute of Ethics Journal* 20 (1): 51–73.

Byrd, W. Carson, Rachelle J. Brunn-Bevel, and Sarah M. Ovink (eds.). 2019. *Intersectionality and higher education: Identity and inequality on college campuses*. New Brunswick: Rutgers University Press.

Carlisle, Liz, Maywa Montenegro de Wit, Marcia S. DeLonge, Alastair Iles, Adam Calo, Christy Getz, Joanna Ory, Katherine Munden-Dixon, Ryan Galt, and Brett Melone. 2019. Transitioning to sustainable agriculture requires growing and sustaining an ecologically skilled workforce. *Frontiers in Sustainable Food Systems* 3: 96.

Ceva, Emanuela, and Maria Paola Ferretti. 2018. The ethics of anti-corruption policies. In *The Routledge Handbook of ethics and public policy*, ed. Annabelle Lever and Andrei Poama, 255–266. Oxon and New York: Routledge.

Cohen, Ronald L. 2016. Restorative justice. In *Handbook of social justice theory and research*, ed. Clara Sabbagh and Manfred Schmitt, 257–272. New York: Springer.

Conde Gutiérrez, Carlos. 2018. Una aproximación a la propiedad intelectual y el acceso a recursos genéticos desde la perspectiva del contrato social y la justicia global. In *Propiedad intelectual. Fundamento y crítica*, ed. Martín Hevía and Facundo M. Rojo, 145–198. Bogotá: Universidad del Externado.

Coolsaet, Brendan. 2016. Towards an agroecology of knowledges: Recognition, cognitive justice and farmers' autonomy in France. *Journal of Rural Studies* 47: 165–171.

De Jonge, Bram. 2011. What is fair and equitable benefit-sharing? *Journal of Agricultural and Environmental Ethics* 24 (2): 127–146.

De Schutter, Olivier. 2009. *International trade in agriculture and the right to food*. Geneva: Friedrich-Ebert-Stiftung.

Deplazes-Zemp, Anna. 2019. Challenges of justice in the context of plant genetic resources. *Frontiers in plant science* 10: 1266.

Desai, Pranav N. 2007. Traditional knowledge and intellectual property protection: Past and future. *Science and Public Policy* 34 (3): 185–197.

Dodds, Walter K. 2005. The commons, game theory and aspects of human nature that may allow conservation of global resources. *Environmental Values* 14: 411–425. https://doi.org/10.3197/096327105774462683.

Drewery, Wendy. 2016. Restorative practice in New Zealand schools: Social development through relational justice. *Educational Philosophy and Theory* 48 (2): 191–203.

El Mujtar, Verónica, Nacira Muñoz, Barbara Prack McCormick, Mirjam Pulleman, and Pablo Tittonell. 2019. Role and management of soil biodiversity for food security and nutrition; where do we stand? *Global Food Security* 20: 132–144.

Figueroa, Robert Melchior. 2011. Indigenous peoples and cultural losses. In *The Oxford handbook of climate change and society*, ed. John S. Dryzek, Richard B. Norgaard, and David Schlosberg, 232–250. Oxford: Oxford University Press.

Fraser, Nancy. 1998. Social Justice in the Age of Identity Politics: Redistribution, Recognition, and Participation. In *The Tanner Lectures of Human Values*, ed. Grethe B. Peterson, 1–67. Salt Lake City: The University of Utah Press.

Fricker, Miranda. 2007. *Epistemic injustice: Power and the ethics of knowing*. Oxford and New York: Oxford University Press.

Geertsema, Willemien, Walter A.H. Rossing, Douglas A. Landis, Felix J.J.A. Bianchi, Paul C.J. Rijn, Joop H.J. Schaminée, Teja Tscharntke, and Wopke Werf. 2016. Actionable knowledge for ecological intensification of agriculture. *Frontiers in Ecology and the Environment* 14 (4): 209–216.

Gomberg, Paul. 2007. *How to make opportunity equal*. New York: Wiley Blackwell.

Grey, Sam. 2019. Returning to the source: Revisiting Arendtian forgiveness in the politics of reconciliation. *Theoria* 66 (161): 37–65.

Gupta, Anil K. 2006. From sink to source: The Honey Bee Network documents indigenous knowledge and innovations in India. *Innovations* (summer): 49–66.

Gupta, Anil K. 2010. Grassroots green innovations for inclusive, sustainable development. In *The Innovation for Development Report 2009–2010, Strengthening Innovation for the Prosperity of the Nations*, ed. Augusto Lopez-Claros, 137–146. Houndmills and New York: Palgrave Macmillan.

Hamilton, Chris. 2008. Intellectual property rights, the bioeconomy and the challenge of biopiracy. *Life Sciences Society and Policy* 4 (3): 26.

Hassemer, Winfried. 2009. *Warum Strafe sein muss: Ein Plädoyer*. Berlin: Ullstein.

Hassoun, Nicole. 2009. Free trade and the environment. *Environmental Ethics* 31 (1): 51–66.

Heins, Volker. 2009. The place of property in the politics of recognition. *Constellations* 16 (4): 579–592.

Herrera Vásquez, Sandra, and Eduardo Rodríguez Yunta. 2004. Etnoconocimiento en Latinoamérica: apropiación de recursos genéticos y bioética. *Acta bioethica* 10 (2): 181–190.

Hickel, Jason. 2016. The true extent of global poverty and hunger: Questioning the good news narrative of the millennium development goals. *Third World Quarterly* 37 (5): 749–767.

Ickowitz, Amy, Bronwen Powell, Dominic Rowland, Andrew Jones, and Terence Sunderland. 2019. Agricultural intensification, dietary diversity, and markets in the global food security narrative. *Global Food Security* 20: 9–16.

Iliescu, Adrian-Paul, Ileana Dascălu, Thierry Ngosso, and Naomi van Steenbergen. 2018. Intergenerational justice in the context of developing countries. In *Towards the Ethics of a Green Future*, ed. Marcus Düwell, Gerhard Bos, and Naomi van Steenbergen, 110–130. London: Routledge.

Jamieson, Dale. 1994. Global environmental justice. *Royal Institute of Philosophy Supplements* 36: 199–210.

Jefferson, Richard A. 2006. Science as social enterprise: The CAMBIA BiOS initiative. *Innovations: Technology, Governance, Globalization* 1 (4): 13–44.

King, Martin Luther. 1963. "I Have a Dream," address delivered at the March on Washington for jobs and freedom. The Martin Luther King, Jr. Research and Education Institute. https://kinginstitute.stanford.edu/king-papers/documents/i-have-dream-address-delivered-march-washington-jobs-and-freedom. Accessed Dec 2019.

Kloppenburg, Jack. 2005. *First the seed: The political economy of plant biotechnology*, 2nd ed. Madison: University of Wisconsin Press.

Koepsell, David. 2016. *Scientific integrity and research ethics: An approach from the ethos of science*. Cham: Springer.

Korthals, Michiel. 2018. Is intensive farming ethically acceptable? *Annals of Advanced Agricultural Sciences* 2 (2): 15–29.

Kroma, Margaret M., and Cornelia Butler Flora. 2003. Greening pesticides: A historical analysis of the social construction of farm chemical advertisements. *Agriculture and Human Values* 20 (1): 21–35.

Langat, Pinky, Dmitri Pisartchik, Diego Silva, Carrie Bernard, Kolby Olsen, Maxwell Smith, Sachin Sahni, and Ross Upshur. 2011. Is there a duty to share? Ethics of sharing research data in the context of public health emergencies. *Public Health Ethics* 4 (1): 4–11.

Lappé, Frances Moore, Jennifer Clapp, Molly Anderson, Robin Broad, Ellen Messer, Thomas Pogge, and Timothy Wise. 2013. How we count hunger matters. *Ethics & International Affairs* 27 (03): 251–259.

Loo, Clement. 2014. Towards a more participative definition of food justice. *Journal of Agricultural and Environmental Ethics* 27 (5): 787–809.

Marion Suiseeya, Kimberly R. 2014. Negotiating the Nagoya protocol: Indigenous demands for justice. *Global Environmental Politics* 14 (3): 102–124.

McGeer, Victoria, and Philip Pettit. 2015. The desirability and feasibility of restorative justice. *Raisons Politiques* 59 (3): 17–33.

McIntyre, Beverly D., Hans R. Herren, Judi Wakhungu, and Robert T. Watson. 2009. *International assessment of agricultural knowledge, science and technology for development (IAASTD): Synthesis report with executive summary: A synthesis of the global and sub-global IAASTD reports*. Washington, DC: Island Press.

McMichael, Anthony. 2017. *Climate change and the health of nations: Famines, fevers, and the fate of populations*. Oxford: Oxford University Press.

Merton, Robert K. 1942. Science and technology in a democratic order. *Journal of Legal and Political Sociology* 1 (1–2): 115–126.

Metz, Thaddeus. 2009. African moral theory and public governance: Nepotism, preferential hiring and other partiality. In *African ethics: An anthology for comparative and applied ethics*, ed. F. Munyaradzi, 335–356. Pietermaritzburg: University of KwaZulu-Natal Press.

Miller, David. 2017. Justice. In *The Stanford Encyclopedia of philosophy* ed. Edward N. Zalta. Stanford: Stanford University.

Munzer, Stephen R. 2012. Corrective justice and intellectual property rights in traditional knolwedge. In *New frontiers in the philosophy of intellectual property*, ed. Annabelle Lever, 58–87. Cambridge: Cambridge University Press.

Neira, Hermán, Lorena Inés Russo, and Bernardita Álvarez Subiabre. 2019. Ecocidio. *Revista de Filosofía* 76: 127–148.

Ooms, Gorik. 2010. Why the West is perceived as being unworthy of cooperation. *Journal of Law, Medicine and Ethics* 38 (3): 594–613. https://doi.org/10.1111/j.1748-720X.2010.00514.x.

Oteros-Rozas, Elisa, Adriana Ruiz-Almeida, Mateo Aguado, José A. González, and Marta G. Rivera-Ferre. 2019. A social–ecological analysis of the global agrifood system. *Proceedings of the National Academy of Sciences*. https://doi.org/10.1073/pnas.1912710116.

Patel, Raj. 2013. The long green revolution. *The Journal of Peasant Studies* 40 (1): 1–63.

Poama, Andrei. 2015. Restorative justice: The institutional turn. *Raisons politiques* 59 (3): 7–16.

Robaey, Zoë. 2016. Gone with the wind: Conceiving of moral responsibility in the case of GMO contamination. *Science and Engineering Ethics* 22 (3): 889–906.

Robinson, Daniel. 2008. Beyond 'protection': Promoting traditional knowledge systems in Thailand. In *Patenting lives: Life patents, culture and development*, ed. Johanna Gibson, 101–130. Aldershot: Ashgate Pub.

Rozzi, Ricardo. 2012. Biocultural ethics: Recovering the vital links between the inhabitants, their habits, and habitats. *Environmental Ethics* 34 (1): 27–50.

Schneider, Ingrid. 2010. *Das Europäische Patentsystem. Wandel von Governance durch Parlamente und Zivilgesellschaft*. Frankfurt am Main and New York: Campus.

Schroeder, Doris, and Balakrishna Pisupati. 2010. *Ethics, justice and the convention on biological diversity*. Nairobi: United Nations Environmental Program.

Schweiger, Gottfried. 2014. Recognition theory and global poverty. *Journal of Global Ethics* 10 (3): 267–273.

Shiva, Vandana, and Poonam Pandey. 2006. *Biodiversity based organic farming: A new paradigm for food security and food safety*. New Dehli: Navdanya.

Solum, Lawrence B. 2004. Procedural justice. *Southern California Law Review* 78: 181–321.

Stephan, Paula E. 2012. *How economics shapes science*. Cambridge, MA: Harvard University Press.

Sterckx, Sigrid. 2005. The ethics of patenting: Uneasy justifications. In *Death of patents*, ed. Peter Drahos, 175–211. Oxford: Lawtext Publishing.

Stojanovic, Milutin. 2019. Biomimicry in Agriculture: Is the ecological system-design model the future agricultural paradigm? *Journal of Agricultural and Environmental Ethics* 32 (5–6): 789–804.

Thomas, Gigesh, and Johan De Tavernier. 2017. Farmer-suicide in India: Debating the role of biotechnology. *Life Sciences, Society and Policy* 13 (1): 8.

Thompson, Paul B. 2017. *The spirit of the soil: Agriculture and environmental ethics*. New York: Routledge.

Timmermann, Cristian. 2017. Harvesting the uncollected fruits of other people's intellectual labour. *Acta bioethica* 23 (2): 259–269. https://doi.org/10.4067/s1726-569x2017000200259.

Timmermann, Cristian, and Georges F. Félix. 2019. Ethical issues involving long-term land leases: A soil sciences perspective. In *Sustainable governance and management of food systems*, ed. Eija Vinnari and Markus Vinnari, 287–292. Wageningen: Wageningen Academic Publishers.

Timmermann, Cristian, and Henk van den Belt. 2012. Global justice considerations for a proposed "Climate Impact Fund" . *Public Reason* 4 (1–2): 182–196.

Tittonell, Pablo. 2013. *Farming systems ecology: Towards ecological intensification of world agriculture*. Wageningen: Wageningen Universiteit.

Tittonell, Pablo, Laurens Klerkx, Frederic Baudron, Georges F. Félix, Andrea Ruggia, Dirk van Apeldoorn, Santiago Dogliotti, Paul Mapfumo, and Walter AH. Rossing. 2016. Ecological intensification: Local innovation to address global challenges. *Sustainable Agriculture Reviews* 19: 1–34.

van der Ploeg, Jan Douwe. 2014. Peasant-driven agricultural growth and food sovereignty. *Journal of Peasant Studies* 41 (6): 999–1030.

van Dijke, Marius, and David De Cremer. 2016. Justice in the work setting. In *Handbook of social justice theory and research*, ed. Clara Sabbagh and Manfred Schmitt, 315–332. New York: Springer.

Vanloqueren, Gaëtan, and Philippe V. Baret. 2009. How agricultural research systems shape a technological regime that develops genetic engineering but locks out agroecological innovations. *Research Policy* 38 (6): 971–983.

Villarroel, Raúl. 2013. Ética del desarrollo, democracia deliberativa y ciudadanía ambiental: El desafío global de la sustentabilidad. *Acta bioethica* 19 (2): 189–198.

Waldmueller, Johannes. 2015. Agriculture, knowledge and the 'colonial matrix of power': Approaching sustainabilities from the Global South. *Journal of Global Ethics* 11 (3): 294–302.

Zhang, Shuwei. 2017. Social justice, institutional trust and public cooperation intention. *Acta Psychologica Sinica* 49 (6): 794–813.

Chapter 10
Feasibility and Justice: The Need for Diverse Innovation Streams

Abstract This chapter argues that a first major step towards changing how natural and social environments are treated is to counter the tolerance of short-term extractivism by linking our innate desires to build a home with long-term commitments to the continuity of biocultures. To establish social and environmental sustainability, the six-dimensional social justice perspective is treated as the baseline for the future design of agricultural innovation policies. After applying this framework to three current problem fields, the chapter discusses possible commitments to political feasibility and their limitations, and concludes by suggesting some research lines for future work.

Keywords Political feasibility · Stewardship · Care ethics · Relational justice · Future of farming

10.1 Back to a Basic Principle: Farming Like We Are Here to Stay

In 2008, the Cuban agronomist Fernando Funes-Monzote submitted a doctoral thesis to Wageningen University with the title "Farming like we are here to stay". This title reveals that we have abandoned long-term commitments to maintaining farming systems. Writing about an island that has suffered the consequences of centuries-long colonialism, the author highlights a problem that is deeply anchored in Latin America and other regions that have been treated as colonies for the rapid extraction of profits: the lack of long-term planning and commitment to the land, its people and nature. While some of the people who moved to the so-called New World arrived with a deep commitment to build a new home and set roots, many others arrived with the goal of making as much money as quickly as possible, often by any means necessary. This way of doing business was so widely propagated that it even spawned a Spanish saying that has fallen into disuse: "*para hacerse la América*" [to make oneself the Americas]. The chance to earn massive revenues in a short timespan came at a gigantic cost for the local population and the environment. Whole landscapes were changed by concentrating on the production of single cash crops, bringing rapid

C. Timmermann, *Social Justice and Agricultural Innovation*, The International Library of Environmental, Agricultural and Food Ethics 31,
https://doi.org/10.1007/978-3-030-56193-2_10

desertification and deforestation as a consequence, as the cases of the north-east of Brazil and Madeira exemplify. Whole peoples where exploited up to a level of near extinction, as was the case with the indigenous population of Haiti (Patel and Moore 2017), while others became extremely vulnerable to diseases and halts in physiological development due to the enormous growth in malnutrition that resulted from the lack of agricultural diversification (de Castro 1961). Maintaining this mentality was partly made possible by never understanding the worked land as a home (Berry 1977/2015)—the true home was located somewhere else: mostly in Europe. This often led to the belief that one would never have to suffer, in old age, the consequences of careless stewardship and one's hostile relation to the natural and social environment. In other cases, a shorter life was willingly accepted as the price of a life of extravagance.

Nowadays, looking at our food production systems, we are left to ask ourselves: how much of this near-sightedness can be observed in today's agriculture, and what role has innovation played? It is a historic irony that the very systems that allowed humanity to settle in a particular piece of land have evolved, in so many places around the world, into a corporate machinery that depletes and erodes the very soils that they were meant to maintain, and is displacing so many people by threatening the livelihood of smallholders (Mazoyer and Roudart 2006).

To make sure that agricultural innovation systems address the many social, economic and environmental sustainability issues of agriculture, we need to align them to meet the many demands of social justice that I have addressed throughout this book. While the range of problems to be solved is huge, so are the opportunities to address the demands of social justice and thereby motivate a much larger and more diverse group of people to regain trust in innovation and contribute to common research projects.

10.2 Realizing Justice: Identifying Conflicts and Opportunities Through a Social Justice Approach

Let us recapitulate. What can the six-dimensional social justice framework presented here do for the assessment of agricultural innovation? This assessment involves three steps (see Fig. 10.1): we start by examining what the four forms of social justice demand; after that we introduce the possible measures that would allow these demands to be realized; and we end by discussing possible loose ends that will have to be addressed afterwards in reconciliatory processes.

I will now show, with three examples, the potential that the framework introduced here has for innovation assessment: (i) the introduction of genetically modified (GM) crops, (ii) the problem of "whitesplaining" among members of international research groups, and (iii) the displacement of horticultural landraces.

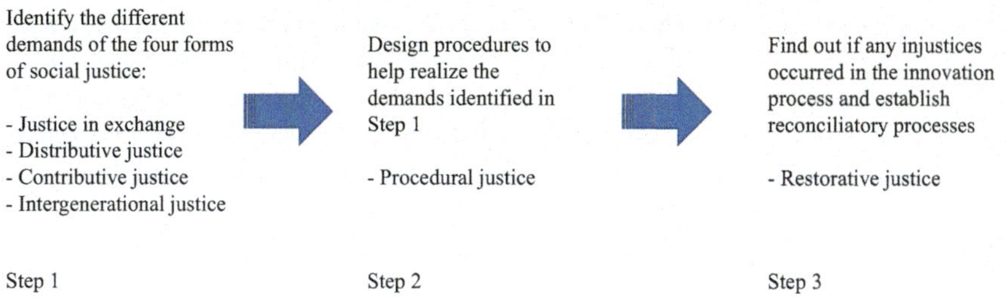

Fig. 10.1 Three steps to realize the demands of social justice

10.2.1 A Social Justice Perspective on the Introduction of GM Crops

The introduction of GM crops has triggered a highly polarized discourse that has made it very difficult to reach a consensus. We have witnessed arguments that accuse biotechnology firms of playing God and putting a price tag on the sanctity of life (van den Belt 2003, 2009), which were later displaced from the discussion in favour of concentrating on issues of safety (Hicks 2017), while at the same time opponents of GM crops are seen as privileged people who, due to unfounded fears, are hindering solutions to the huge welfare burden of global hunger (Borlaug 2000). As each group appeals to a highly morally-laden cause—the sanctity of life, biosafety and the need to feed the poor—the opposing party often comes to be regarded as a group with whom one cannot constructively engage in discussions, and in extreme cases, an adversary with whom one should not even associate. This extreme polarization does not allow the two main groups to work together towards establishing policies that would lead to a win–win situation (Jauernig et al. 2019), and impedes taking even the smallest steps towards improving current practices. This polarization also concentrates the whole discussion on addressing the two extreme positions, at the expense of not identifying or dialoguing about other pressing issues (Hicks 2017).

From a social justice perspective, what are the other conflicts and opportunities that we are missing? Let us briefly revise the six dimensions:

Justice in exchange. GM crop developers and distributors should refrain from making unrealistic or false claims, and make sure that farmers are aware of both the full benefits and costs of the technology.

Distributive justice. A worry here is that research on GM crops has focused primarily on cash crops for animal feed, biofuels and the textile industry, and to a much lower extent on addressing the direct food needs of people in tropical regions by improving local varieties.

Contributive justice. Extensive use of intellectual property has made innovation processes more exclusive and has made it difficult to carry out independent trials.

Intergenerational justice. The use of GM crops should not negatively affect biodiversity or lead to soil erosion.

Procedural justice. There needs to be an informed public deliberation on the benefits and dangers of the technology. The public should have a right to object to the use of GM crops. Policy-makers and biotechnology companies need to work hand in hand to make sure that if the technology is used, it is used in a socially and environmentally sound manner, for example, by obliging farmers through contracts to maintain refuge areas for biodiversity and the provision of ecosystem services.

Restorative justice. Biotechnology companies and distributors need to engage in reconciliatory processes with farmers who were deceived or improperly informed and have suffered negative consequences, eventually offering their expertise to help them recover. The companies need to establish internal policies to encourage public outreach before introducing new technologies.

By being invited to assess technologies through this multidimensional approach, we can systematically identify more points of conflict and interaction that allow us to work progressively towards the different solutions that the responsible introduction and withdrawal of technologies requires. The lack of consultation and failure to provide adequate information have led to a situation where farmers feel that a technology has been imposed on them, even when it is not suitable for local conditions (Nyéléni Forum for Food Sovereignty 2007). Now a restorative process is needed to address this injustice.

10.2.2 Whitesplaining and Patronizing Researchers

Another practice that is increasingly condemned in international collaborations is referred to as "whitesplaining". By borrowing this concept from feminist studies, which condemn as "mansplaining" the patronizing interruptions or reformulations made by male peers, a wide range of people have voiced opposition to similar behaviour by researchers from the western Global North, mostly directed at scientists from the Global South. Particularly when such practices do not help to clarify arguments, and are only targeted at a select group of peers, they may be perceived as discriminatory and a nuisance, not only for the people who originally express a particular finding or observation, but also for those who want to hear it from the original source. Such practices are particularly harmful for science if partners in the Global North end up becoming the principal voice concerning a subject on which they have less or no epistemic authority. Let us visualize the different social justice issues that come into play with this practice.

Justice in exchange. This practice may lead to researchers from the Global South not being identifiable as the original source of a new idea or important observation. This may lead to lost opportunities for future rewards.

Distributive justice. This type of interruption goes against the fair allocation of discussion time and diverts attention away from those who were originally entitled to it.

There is also a risk that the remark is decontextualized and attention is drawn away from the main problem.

Contributive justice. These practices, particularly when they are repeated, discourage people from engaging in discourse with people from outside one's community, and impede participation as peers.

Intergenerational justice. Failing to address such practices leads, in the long run, to historical discrimination and a lack of role models with whom one shares a cultural background.

Procedural justice. Educational campaigns are needed on the harmful effects of such practices, and special training is needed for chairs of such discussions to be good arbiters.

Restorative justice. Reconciliation processes are needed, where everyone is obliged to share their perspective and justify it.

Here again, by taking a wider social justice perspective, we can identify a wider set of negative consequences. This allows us to have much stronger arguments than qualifying such practices as rude. Employing a language of justice, we can condemn the hoarding of attention as unjustly taking advantage. This perspective not only condemns those who unjustly take advantage, but also those in the audience who fail to repudiate such practices.

10.2.3 Displacement of Horticultural Landraces

Also of concern is when landraces around home gardens are quickly displaced in favour of improved varieties or specialization in key staple crops. A careless displacement of landraces may have a major disempowering effect on local women, as they are the ones who overwhelmingly select and improve the varieties that are cultivated near homes (Fadda 2016). In many regions, depriving local women of the opportunity to engage in crop improvement and seed exchange takes away one of the few opportunities they may have to interact with other women outside their household and share experiences, observations and ideas, and get feedback and help (Shiva and Pandey 2006). Technocratic advice in such cases is regularly seen as superior to any local seed improvement capabilities or garden design. Here too, a social justice framework can reveal additional worries.

Justice in exchange. Informed consent on the acquisition of local varieties needs to make local women aware of the lost freedoms and of the position of unilateral dependency into which they will enter, so as to allow them to weigh the full benefits and costs of using improved varieties.

Distributive justice. Local women need to be aware that, by acquiring improved seeds from industry, they will lose their liberty to focus their attention on improving the varieties they consider more important and choosing desirable traits.

Contributive justice. Acquiring seeds from outsiders may have a disempowering effect. Local women will have to carefully evaluate whether the freed time they may gain by acquiring improved varieties can be used at their discretion, and whether eventual increases in yields leave them better off overall.

Intergenerational justice. Special care needs to be taken to avoid the loss of heritage varieties and to secure continuous access to them in the future. Attention needs be paid to ensure that future generations do not inherit oppressive structures of dependency.

Procedural justice. Local women also need to have a voice in farm design and the choice of seed varieties and traits.

Restorative justice. Reconciliatory processes need to be established in cases of major losses of freedom and undesirable states of dependency. Eventually, reempowerment must be sought.

A social justice lens reveals that a narrow focus on improving yields in home gardens may come at the cost of disempowerment and place local women in an unilateral position of dependency (Preston and Wickson 2016). Special care needs to be taken to ensure that the added advantages of improved yields compensate the social and individual disadvantages of losing the capability to select, exchange and improve seeds.

10.3 Working Towards a Wider Agricultural Transition

Throughout this book we have seen that different innovation streams vary significantly in their effect on the human and natural environment. In terms of justice, we can observe that non-proprietary science systems, especially agroecology, show the greatest compatibility with the demands of all four forms of social justice. These systems avoid the worst problems of the commercialization of research, by not leaving people without access merely for lacking sufficient monetary resources, and by not over-incentivizing research for the richer markets. Such farming methods are more participatory when it comes to their optimization, as they are free from intellectual property constraints. Treating knowledge as a commons avoids many of the conflicts in scientific enterprises that are triggered by strong market incentives that tempt researchers to withhold sufficient credit from intellectual contributors. Lastly, these food systems are more socially and environmentally sustainable, as there are fewer incentives to sell and use innovations beyond ecologically responsible thresholds.

Yet, in terms of practicality, food systems that are labour intensive, such as agroecology, hit a major feasibility limitation: they require enough people who are willing to work the land by choice and not only by necessity (Dumont et al. 2016). Furthermore, they require people to work the land with sufficient devotion to acquire comprehensive knowledge and develop strong ties to the land, leading to good stewardship (Timmermann and Félix 2015). In relation to this problem, we also need to take into consideration that younger people in particular are moving to the cities, leaving insufficient people in rural areas to work the land (Bernstein 2014). The technological feasibility of further improving automation makes a future scenario where food can be grown effortlessly particularly attractive. Additionally, to reduce food miles, more food needs to be grown in urban areas, in particular rapidly perishing vegetables, which also raises questions about the choice and mixture of farming systems for the cities (Rydin et al. 2012).

Continuous advancement in the mechanisation of agriculture makes us wonder about the role and nature of agricultural work (Berry 1977/2015). A world where food is grown without effort is also a world where less work is available, and there are substantial concerns about a future where technologization has made human work redundant (cf. Steinvorth 2009; Danaher 2019). This raises complex questions about how to understand humanity without the need to work (Bottazzi 2019). It also makes it necessary for us to expand traditional ideas of benefiting from scientific advancement with notions of being entitled to the advantages of increased industrialization. If work is the key means to secure one's food entitlement (Sen 1981), then we will have to guarantee each person's access to food through other means as work becomes increasingly scarce.

In terms of overall agricultural innovation, despite the added benefits of non-proprietary science systems in terms of participation, we need to be somewhat more critical of their capacity to develop innovations until they are ripe for the market or sufficiently developed to encourage wide adoption. We must remember that one of the motivations for supporting legislation such as the Bayh-Dole Act in the United States in 1980, was that major stakeholder groups condemned the fact that research was not transformed into a sufficient number of innovations. Non-financial motivations prior to that date did not consolidate to provide wide-scale solutions to global hunger. While there are substantial worries about its true level of success, the Green Revolution has increased the aggregate supply of grain, while failing to sufficiently address issues of dietary diversity and food entitlement (Patel 2013). As the current number of hungry and malnourished people clearly reveals, market solutions alone have not been able to fix the world's food problem. Market incentives also encourage the overprovision of calories, and make it extreme lucrative to sell processed food with poor nutritional content, leading to obesity and malnutrition (Korthals 2018), and often to ignore socially and environmentally sustainable practices (Folke et al. 2011; Carolan 2018).

Following Reiss and Kitcher (2009) conclusions on the problem of neglected diseases, it also seems that agriculture needs a system that allows us to harvest efforts made out of different types of motivation, be them financial, out of solidarity, scientific curiosity or the desire to live in a just world. For this, we need to make room for the coexistence of different innovation systems that permit people to participate with different motivations and perspectives, with a willingness to develop new technologies, work on prevention systems, and revise methods and solutions. Nonetheless, we must insist on relying on a normative framework, like the one introduced here, which can connect such different streams of innovation and motivations with achieving socially and environmentally desirable outcomes, by taking into systematic consideration that our individual actions and omissions may, in aggregate, have unwanted consequences or even unforeseen welcomed effects.

In this book I have concentrated on setting out the different social justice demands regarding agricultural innovation that must be met for us to say that we live in a just world. The understanding of justice language that I have offered here considerably expands expectations of what we owe each other. Seeking justice in agricultural innovation, as is the case with the environment, requires us to extend the responsibilities that traditionally concentrate around the State to all people capable of making changes. It is how people interact, use and ignore knowledge and biodiversity, and produce and consume goods, that in aggregate has a huge impact on the environment (Jamieson 1994). Applying justice concepts to specific areas, as is also the case with environmental justice and climate justice, departs from the perspective of ideals of justice, to examine how justice and injustices are experienced (Schlosberg 2013). This may vastly increase citizens' discontent with current practices. For instance, gaining awareness of the multiple demands of intergenerational justice may give rise to a strong political demand to not have to contribute, as individual consumers, to the unsustainable use of resources at the cost of future generations (Timmermann et al. 2018). At the same time, increased participation and development of skills empowers citizens to become agents of justice and thus capable of contributing towards the realization of social justice.

At the same time, a broader understanding of justice inevitably increases the consumer's responsibility for unjust outcomes. The current lack of cooperation of the general population, particularly the wealthy, in reducing their environmental footprint has reached catastrophic levels, in terms of mass extinctions of species, climate collapse and losses of bioheritage. There are only a few years left to see if humanity manages to curb the most disastrous effects of climate change. In terms of justice, as citizens we can no longer credibly shift the full responsibility of our inaction to the incompetence of our (often elected) governments.

10.4 Change and Political Feasibility

There is a broad consensus that we cannot continue to farm under business-as-usual models (McIntyre et al. 2009). In the highly polarized debate on which food future or futures are to be followed, we need to highlight some of the key factors that need to be taken into consideration before committing to any large-scale reform.

1. *Food is a continuous need.* Any large-scale disruption in the food system is likely to have a massive human cost, and any major change needs to take this risk seriously (De Schutter 2017). As our experience with natural disasters, collapses of economies and political regimes and war tells us, disruptions in the food supply disproportionally affect the poor and the most vulnerable populations.

2. *Consolidation of the food and agriculture sector.* The sector is dominated by large players who have enormous lobbying power and the ability to cause major disruptions in the food supply, by inflating prices, delaying deliveries, and even deliberately spoiling foodstuffs (ETC Group 2008; Korthals 2018). Ignoring their power and interests may come at a significant cost in welfare.
3. *Sustainable alternatives are knowledge-intensive.* A large-scale transition can only be achieved at the speed at which farmers become familiar with a new skillset (Carlisle et al. 2019). Skills take time to learn and teach.
4. *Technology and the amplification of inequalities.* To make sure that people benefit from scientific and technological advancement, we need to support the development of skills to ensure that they can make the most of technologies. Skills and environmental and social resources can greatly amplify the potential of technologies (Glover et al. 2019). Public policies that are meant to improve the situation of the worst off need to include a broader assessment of their development politics.
5. *Cooperation from the non-farming population.* The world has become extremely dependent upon cheap food (Patel and Moore 2017). It will take a major effort to teach people about the true costs of food production and to accept higher food prices. Minimum wages need to be increased so that people can afford food with a lower environmental footprint. A global food system reform will have to go hand in hand with major global political reform.
6. *Ties between nature and culture.* The natural environment is the key element for many cultural understandings and identifications (Rozzi 2013). To put it in terms of capabilities, while humans depend on various ecosystem services for their functioning (Schlosberg 2013), many indigenous communities are dependent upon specific natural environments, including their terrain, flora and fauna, for their functioning as indigenous communities. If we lose a natural environment, then we also lose the cultures that are embedded in that environment (Anthony 2013).

The wide scope of these problems demands a stronger citizens' involvement that goes beyond their active participation in science governance. Aligning agricultural innovation with the demands of social justice will have to go hand in hand with a wider social commitment to agricultural justice. Together with improving government policies and their efficacy, it is crucial for citizens to become responsible and cooperative food consumers (Timmermann et al. 2018). There a two major factors that encourage contributing to a larger socio-environmental transformation:

1. *Awareness, change and momentum.* A key factor for making a major shift feasible is that it is pursued at the right time. Public awareness and willingness to change has a certain momentum such that it can be a major contributing factor for change (Gilabert and Lawford-Smith 2012), but later it becomes more difficult to raise enthusiasm due to apathy or tiredness about the topic. Current social unrest triggered by stagnating wages and increases in living costs show a clear and strong demand for systemic change.
2. *Climate emergency.* At the time of this writing we have already heard warnings about the disastrous levels of greenhouse gas emissions from all scientific sectors.

> In human history, there has never been a greater time of urgency to reshape our food production systems (McMichael 2017).

Reasons to stay optimistic are, however, particularly scarce. Voters of major economies have recently elected heads of state who show no or little respect for the natural environment and evidence-based policies, and have often further eroded welfare state policies protecting the most food vulnerable. Moreover, many countries in the world continue to be ruled by despots. In addition, major corporations continue to fight tighter environmental regulations and labour rights. An upfront confrontation with these powerful actors could lead to massive food shocks in the short run, which will be suffered the most by the global poor. Under current political realities, any transformation of food systems would have to make commitments to political feasibility.

10.5 On Making Compromises

When departing from the ideals of social justice that we have discussed in this book in order to make compromises or even sacrifices for political feasibility, we need to take into account that there are still three major separate (but partially overlapping) argumentation lines to prevent a too drastic deviation from the ideal: human rights, sustainability principles and Sustainable Development Goals.

In the case of human rights, we need to take special care not to take the extensive Universal Declaration of 1948 and all the rights it aims to secure for granted. The catastrophic events of World War II encouraged nations from all over the world to make extensive commitments and it is very difficult to achieve such goodwill, other than in cases of extreme events (Morsink 1999). Here we also need to take care that alien elements are not too easily introduced and placed on a par with the original set of rights, as is the case with the proposed human right to intellectual property, without specifying the scope and limitations of such rights (cf. UN Committee on Economic Social and Cultural Rights 2006).

The major political influence of social movements based on sustainability principles is another force that has a strong potential to reduce our environmental footprint in view of the interests and survival of future generations. The broad political consensus on such principles was an achievement that we should also not undermine.

Civil society efforts to push governments to work towards achieving the Sustainable Development Goals are also becoming a strong force. In a handful of countries, governments are becoming increasingly liable to make empty promises with regard to sustainability (The Associated Press 2019). The destabilizing factors of continuing, as innovators and politicians, to not fulfil promises, and of failing to meet political targets, need to be taken increasingly seriously.

After gaining wider awareness of the many social justice issues that agricultural innovation involves, we also need to acknowledge the price paid by scientific communities, farmers and society at large for failing to do what justice requires. Distrust is the confident belief that a particular group or institution will not act as justice demands (Krishnamurthy 2015). If scientists and professional innovators are grouped into a homogeneous category that is deemed untrustworthy by large segments of the population, then we will have a situation where crucial cooperation opportunities are insufficiently explored, or not explored at all, and where advice is rarely sought and often mistrusted. Distrust towards scientists and professional innovators is an outcome that harms society as a whole. It leads to a situation where voters are less supportive of pro-science political agendas and there is a lower demand for evidence-based policies. For scientists and innovators this means less opportunities for work and for having a positive social and environmental impact.

Establishing trust in innovation systems requires these systems to comply with the demands of justice. As such, trust has a public good character: everyone who benefits from such a system has the moral obligation to work towards establishing this good. This requires cooperation in complying with rules, working towards transparency, developing skills and inclusion, monitoring each other, and participating in science governance. The state, innovators and the wider citizenship are thus obliged to do their respective shares in establishing and maintaining this important public good.

Considering these issues, I cannot recommend sacrificing key social justice demands in favour of political feasibility below a certain threshold. Full capitulation of single demands may have a disastrous effect on taking human rights seriously as a complex set of political, civil, economic, social and cultural demands in their entirety. In addition, technological advancement should have made the progressive realization of social and environmental demands easier. Large-scale social discontent and rebellion in response to a disproportional raise in living costs and a regression towards longer weekly working hours comes as no surprise. Such trends are incompatible with a scenario where people truly benefit from scientific advancement. If currents trends continue, society may even create a disincentive to contribute to scientific advancement, particularly when such contributions lead to more corporate capture and increased rent-seeking activities, with no or insufficient improvement of social welfare and environmental protection.

It is urgent for the private and public sectors to concentrate on exploring win–win situations, where addressing the demands of social justice leads to better results for innovation, society and the environment, such as the conservation of agrobiodiversity to facilitate future plant breeding and linking soil restoration with carbon capture. Institutions will need to reconsider the moral relevance of the divide between positive and negative duties (or between acts and omissions). In many cases, the building of institutions for cooperation, reconciliation and problem-solving (a positive duty) may come significantly cheaper than a strict commitment to avoid harm (a negative duty). In some situations, building these institutions may ultimately lead to better outcomes for all involved. These strategies need to be assessed on a case-by-case basis, in view of the full set of demands of social justice and the counterfactual situation where no agreement was reached.

10.6 Future Work

Three major research lines will have to be left for future work: empirical studies on the relevance of this framework, relational justice and care ethics, and engaging in a social dialogue on the future of farming.

10.6.1 Future Avenues for Empirical Studies

A philosophical examination helps us to identify the many demands of justice, yet it is beyond its limits to gather empirical data about the actual importance that people give to these demands. It is crucial to find out about these, as we will have to assess to what extent, from a social justice perspective, we can prioritize certain areas to which people give a higher weighting, and it allows us to identify which dimensions of social justice are being neglected, but should not be sacrificed, as is the case with the interests of future generations. Such empirical studies are crucial for identifying which dimensions of social justice are mostly perceived to be unaddressed, and which dimensions are widely ignored as problematic. It also allows us to design educational campaigns to help people to recognize their rights and comply with their duties. These are vital for creating willingness to cooperate and rebuilding trust in science. They are also crucial for addressing the problem of adaptive preferences, where historically discriminated groups adjust their preferences to a lower standard than that to which they would be legitimately entitled.

The design of both ex ante and ex post processes, following principles of procedural justice and restorative justice, needs to consider the empirical evidence, listen to different voices and narratives, build ties to the ideals of social justice, and be continuously optimized.

10.6.2 Relational Justice and Care Ethics

Many of the demands of relational justice, as understood by Iris Young and those inspired by her work, have been discussed under the four forms of justice examined in this book, particularly our obligations to remedy structural injustices. There is, however, a component that has hardly been addressed at all: the recognition of the role and intrinsic value of relations of care. Relations of care to crops, animals and soils—in terms of good husbandry—have been part of agriculture from its very beginning. Agriculture would not have been possible at all without such long-term commitments and an ability to identify the needs of crops, soils and animals, and react and provide accordingly. Here we can also observe other important parameters that are frequently overlooked, such as the particularity of relationships and their context, the different forms of vulnerability and dependency, in particular linked dependencies, and even more generally, the affective dimension (Preston and Wickson 2016). Incorporating these perspectives and dimensions is crucial for any future sensitive social justice assessment. Here it must be noted that the importance of care relations has not only been highlighted in feminist theory, but is also central to many non-Western philosophies (Mackey and Claudie 2015). It remains an open question, to which I am positively inclined, to what extent agricultural innovation should maintain room to exercise such relations and not make them redundant.

10.6.3 The Need for a Social Dialogue on Technology and the Future of Farming

The widespread failure of conventional agricultural innovation to address the demands of social justice demands urgent work. We need to counteract the negative effects of agriculture on both society and the environment. This requires social cooperation, and to obtain social cooperation we need to improve awareness of the current costs of agriculture. Major public education campaigns need to be implemented and expanded.

The time has come at which we urgently have to decide, as a society, how we should value farm work and being a farmer. As a profession that is well over 11,000 years old, farming plays a fundamental role in human nature and culture (Hilmi 2018; Berry 1977/2015). We may ponder whether we should also treat farming as a special interest, in the sense of protecting it as a way of life, or even as a belief system (as *Weltanschaung*). We may be able to do this, even though farming is different from many protected religious beliefs and indigenous communities' customs, in that it is completely open to anyone (who has the means) to freely become or cease to be a farmer. Yet tradition alone seldom establishes a claim for justice. Many professions have disappeared, for better or worse, often creating new ones.

To decide on the future of farming, we need to consider not only the current state of smallholder farming, but also the many ways in which we could improve working conditions for smallholders (Bragdon 2016). Nowadays, the vast majority of research money is destined for conventional farming, leaving agroecology and the development of tools for smallholders underexplored (Tittonell 2013; Altieri et al. 2012). This is particularly blameworthy due to the many underused opportunities to redesign farming systems so that they mimic natural processes and can thus be safely embedded within adjacent ecosystems (Stojanovic 2019). Any social dialogue on the future of farming needs to take into account that there is actually substantial room to improve the experience and output of smallholder agriculture (Timmermann and Félix 2015).

Another issue of concern that we should not take lightly is the question of whether we are indeed entitled, as a generation, to decide to abandon any commitment to farming as a form of life and deprive future generations of this opportunity. In principle, future generations will be able to reverse this negative decision; but when they do, many local varieties and much tacit knowledge and cultural heritage will be lost.

References

Altieri, Miguel A., Fernando R. Funes-Monzote, and Paulo Petersen. 2012. Agroecologically efficient agricultural systems for smallholder farmers: Contributions to food sovereignty. *Agronomy for Sustainable Development* 32 (1): 1–13.

Anthony, Raymond. 2013. Animistic pragmatism and native ways of knowing: Adaptive strategies for overcoming the struggle for food in the sub-Arctic. *International Journal of Circumpolar Health* 72: 21224.

Bernstein, Henry. 2014. Food sovereignty via the 'peasant way': A sceptical view. *Journal of Peasant Studies* 41 (6): 1031–1063. https://doi.org/10.1080/03066150.2013.852082.

Berry, Wendell. 1977/2015. *The unsettling of America: Culture & agriculture.* Berkeley: Counterpoint Press.

Borlaug, Norman E. 2000. Ending world hunger. The promise of biotechnology and the threat of antiscience zealotry. *Plant Physiology* 124 (2): 487–490.

Bottazzi, Patrick. 2019. Work and social-ecological transitions: A critical review of five contrasting approaches. *Sustainability* 11 (14): 3852.

Bragdon, Susan H. 2016. Reinvigorating the public sector: The case of food security, small-scale farmers, trade and intellectual property rules. *Development* 59 (3–4): 280–291.

Carlisle, Liz, Maywa Montenegro de Wit, Marcia S. DeLonge, Alastair Iles, Adam Calo, Christy Getz, Joanna Ory, Katherine Munden-Dixon, Ryan Galt, and Brett Melone. 2019. Transitioning to sustainable agriculture requires growing and sustaining an ecologically skilled workforce. *Frontiers in Sustainable Food Systems* 3: 96.

Carolan, Michael. 2018. *The real cost of cheap food.* Oxon & New York: Routledge.

Danaher, John. 2019. *Automation and Utopia: Human flourishing in a world without work.* Cambridge, MA: Harvard University Press.

de Castro, Josué. 1961. *Geografía Del Hambre.* Santiago de Chile: Editorial Universitaria.

Dumont, Antoinette M, Gaëtan Vanloqueren, Pierre M Stassart, and Philippe V Baret. 2016. Clarifying the socio-economic dimensions of agroecology: Between principles and practices. *Agroecology and Sustainable Food Systems* 40 (1): 24–47.

ETC Group. 2008. *Who owns nature?: Corporate power and the final frontier in the commodification of life.* Ottawa: ETC Group.

Fadda, Carlo. 2016. "The farmer's role in creating new genetic diversity." In *Farmers' Crop Varieties and Farmers' Rights*, ed. Michael Halewood, 57–70. Oxon & New York: Routledge.

Folke, Carl, Åsa Jansson, Johan Rockström, Per Olsson, Stephen R. Carpenter, F. Stuart Chapin, Anne-Sophie. Crépin, Gretchen Daily, Kjell Danell, and Jonas Ebbesson. 2011. Reconnecting to the biosphere. *Ambio* 40 (7): 719.

Funes-Monzote, Fernando. 2008. "Farming like we're here to stay: The mixed farming alternative for Cuba." PhD thesis, Wagenigen University.

Gilabert, Pablo, and Holly Lawford-Smith. 2012. Political feasibility: A conceptual exploration. *Political Studies* 60 (4): 809–825.

Glover, Dominic, James Sumberg, Giel Ton, Jens Andersson, and Lone Badstue. 2019. "Rethinking technological change in smallholder agriculture." *Outlook on Agriculture* 48 (3): 169–180.

Hicks, Daniel J. 2017. Genetically modified crops, inclusion, and democracy. *Perspectives on Science* 25 (4): 488–520.

Hilmi, Angela. 2018. "Peasant farming as a source of life." *Development* 61: 122–128.

Jamieson, Dale. 1994. Global environmental justice. *Royal Institute of Philosophy Supplements* 36: 199–210.

Jauernig, Johanna, Ingo Pies, Paul B. Thompson, and Vladislav Valentinov. 2019. Theorizing agriculture-society tensions: An ordonomic approach to the agrarian vision. In *Sustainable governance and management of food systems: Ethical perspectives*, ed. Eija Vinnari and Markus Vinnari, 634–657. Wageningen: Wageningen Academic Publishers.

Korthals, Michiel. 2018. Is Intensive farming ethically acceptable? *Annals of Advanced Agricultural Sciences* 2 (2): 15–29.

Krishnamurthy, Meena. 2015. (White) Tyranny and the democratic value of distrust. *The Monist* 98 (4): 391–406.

Mackey, Brendan, and David Claudie. 2015. Points of contact: Integrating traditional and scientific knowledge for biocultural conservation. *Environmental Ethics* 37 (3): 341–357.

Mazoyer, Marcel, and Laurence Roudart. 2006. *A history of world agriculture: From the neolithic age to the current crisis*. New York: Monthy Review Press.

McIntyre, Beverly D., Hans R. Herren, Judi Wakhungu, and Robert T. Watson. 2009. *International assessment of agricultural knowledge, science and technology for development (IAASTD): Synthesis report with executive summary: A synthesis of the global and sub-global IAASTD reports*. Washington, DC: Island Press.

McMichael, Anthony. 2017. *Climate change and the health of nations: Famines, fevers, and the fate of populations*. Oxford: Oxford University Press.

Morsink, Johannes. 1999. *The universal declaration of human rights: Origins, drafting, and intent*. Philadelphia: University of Pennsylvania Press.

Nyéléni Forum for Food Sovereignty. 2007. *Declaration of Nyéléni*. Sélingue: Nyéléni Forum for Food Sovereignty.

Patel, Raj. 2013. The long green revolution. *The Journal of Peasant Studies* 40 (1): 1–63.

Patel, Raj, and Jason W. Moore. 2017. *A history of the world in seven cheap things: A guide to capitalism, nature, and the future of the planet*. Oakland, CA: University of California Press.

Preston, Christopher J., and Fern Wickson. 2016. Broadening the lens for the governance of emerging technologies: Care ethics and agricultural biotechnology. *Technology in Society* 45: 48–57.

Reiss, Julian, and Philip Kitcher. 2009. "Biomedical research, neglected diseases, and well-ordered science." *THEORIA. Revista de Teoría, Historia y Fundamentos de la Ciencia* 24 (3): 263–282.

Rozzi, Ricardo. 2013. "Biocultural Ethics: From Biocultural Homogenization Toward Biocultural Conservation." In *Linking Ecology and Ethics for a Changing World: Values, Philosophy, and Action*, ed. Ricardo Rozzi, S. T. A. Pickett, Clare Palmer, Juan J. Armesto and J. Baird Callicott, 9–32. Dordrecht: Springer.

Rydin, Yvonne, Ana Bleahu, Michael Davies, Julio D. Dávila, Sharon Friel, Giovanni De Grandis, Nora Groce, Pedro C. Hallal, Ian Hamilton, and Philippa Howden-Chapman. 2012. Shaping cities for health: Complexity and the planning of urban environments in the 21st century. *the Lancet* 379 (9831): 2079–2108.

Schlosberg, David. 2013. Theorising environmental justice: The expanding sphere of a discourse. *Environmental Politics* 22 (1): 37–55.

Schutter, De., and Olivier. 2017. The political economy of food systems reform. *European Review of Agricultural Economics* 44 (4): 705–731.

Sen, Amartya. 1981. *Poverty and famines. An essay on entitlement and deprivation*. Oxford & New York: Oxford University Press.

Shiva, Vandana, and Poonam Pandey. 2006. *Biodiversity based organic farming: A new paradigm for food security and food safety*. New Dehli: Navdanya.

Steinvorth, Ulrich. 2009. The right to work and the right to develop one's capabilities. *Analyse & Kritik* 1: 101–113.

Stojanovic, Milutin. 2019. Biomimicry in Agriculture: Is the ecological system-design model the future agricultural paradigm? *Journal of Agricultural and Environmental Ethics* 32 (5–6): 789–804.

The Associated Press. 2019. "Activists cheer victory in landmark Dutch climate case." *The New York Times*. Accessed December 20, 2019. https://www.nytimes.com/aponline/2019/12/20/business/bc-eu-netherlands-climate-case.html.

Timmermann, Cristian, and Georges F. Félix. 2015. Agroecology as a vehicle for contributive justice. *Agriculture and Human Values* 32 (3): 523–538.

Timmermann, Cristian, Georges F. Félix, and Pablo Tittonell. 2018. Food sovereignty and consumer sovereignty: Two antagonistic goals? *Agroecology and Sustainable Food Systems* 42 (3): 274–298. https://doi.org/10.1080/21683565.2017.1359807.

Tittonell, Pablo. 2013. *Farming systems ecology: Towards ecological intensification of world agriculture*. Wageningen: Wageningen Universiteit.

UN Committee on Economic Social and Cultural Rights. 2006. General Comment No. 17: The Right of Everyone to Benefit from the Protection of the Moral and Material Interests Resulting from any Scientific, Literary or Artistic Production of Which He or She is the Author (Art. 15, Para. 1 (c) of the Covenant, E/C.12/GC/17). Geneva: United Nations Economic and Social Council.

van den Belt, Henk. 2003. "Enclosing the genetic commons: biopatenting on a global scale." In *Patente am Leben? Ethische, rechtliche und politische Aspekte der Biopatentierung*, ed. Christoph Baumgartner and Dietmar Mieth, 229–243. Paderborn: mentis.

van den Belt, Henk. 2009. Playing god in Frankenstein's footsteps: Synthetic biology and the meaning of life. *Nanoethics* 3 (3): 257–268. https://doi.org/10.1007/s11569-009-0079-6.

The manufacturer's authorised representative in the EU is Springer Nature Customer Service Centre GmbH, Europaplatz 3, 69115 Heidelberg, Germany. If you have any concerns regarding our products, please contact ProductSafety@springernature.com

Printed and bound by CPI Group (UK) Ltd, Croydon, CR0 4YY
30/09/2026
02225445-0004